ENERGY SCIENCE, ENGINEERING AND TECHNOLOGY

CRITICAL MATERIALS STRATEGY FOR CLEAN ENERGY TECHNOLOGIES

ENERGY SCIENCE, ENGINEERING AND TECHNOLOGY

Additional books in this series can be found on Nova's website under the Series tab.

Additional E-books in this series can be found on Nova's website under the E-books tab.

ENERGY SCIENCE, ENGINEERING AND TECHNOLOGY

CRITICAL MATERIALS STRATEGY FOR CLEAN ENERGY TECHNOLOGIES

NICHOLAS J. BARTLETT
EDITOR

Nova Science Publishers, Inc.
New York

For permission to use material from this book please contact us:
Telephone 631-231-7269; Fax 631-231-8175
Web Site: http://www.novapublishers.com

LIBRARY OF CONGRESS CATALOGING-IN-PUBLICATION DATA

Critical materials strategy for clean energy technologies / editor, Nicholas J. Bartlett.
p. cm.
Includes reports by the Department of Energy and Marc Humphries, Congressional Research Service.
Includes bibliographical references and index.
ISBN 978-1-61209-829-6 (hardcover)
1. Rare earths--Government policy--United States. 2. Clean energy industries--United States--Materials. 3. Energy conservation--United States--Materials. 4. National security--United States. I. Bartlett, Nicholas J. II. Humphries, Marc. III. United States. Dept. of Energy.
TA480.R3C75 2011
333.8'54940973--dc22

2011002525

Published by Nova Science Publishers, Inc., New York

CONTENTS

PREFACE

Each day researchers, entrepreneurs and many others across the United States are working to develop and deploy clean energy technologies that will enhance our security, reduce pollution and promote prosperity. Many new and emerging clean energy technologies, such as the components of wind turbines and electric vehicles, depend on materials with unique properties. The availability of a number of these materials is at risk due to their location, vulnerability to supply disruptions and lack of suitable substitutes. Ensuring reliable access to critical materials will help the United States lead in the new clean energy economy. This new book examines the role of rare earth metals and other materials in the clean energy economy, based on data collected and researched by the U.S. Department of Energy.

Chapter 1- This report examines the role of rare earth metals and other materials in the clean energy economy. It was prepared by the U.S. Department of Energy (DOE) based on data collected and research performed during 2010. Its main conclusions include:

- Several clean energy technologies—including wind turbines, electric vehicles, photovoltaic cells and fluorescent lighting—use materials at risk of supply disruptions in the short term. Those risks will generally decrease in the medium and long term.
- Clean energy technologies currently constitute about 20 percent of global consumption of critical materials. As clean energy technologies are deployed more widely in the decades ahead, their share of global consumption of critical materials will likely grow.
- Of the materials analyzed, five rare earth metals (dysprosium, neodymium, terbium, europium and yttrium), as well as indium, are assessed as most critical in the short term. For this purpose, "criticality" is a measure that combines importance to the clean energy economy and risk of supply disruption.
- Sound policies and strategic investments can reduce the risk of supply disruptions, especially in the medium and long term.
- Data with respect to many of the issues considered in this report are sparse.

Chapter 2- The concentration of production of rare earth elements (REEs) outside the United States raises the important issue of supply vulnerability. REEs are used for new energy technologies and national security applications. Is the United States vulnerable to supply disruptions of REEs? Are these elements essential to U.S. national security and economic well-being?

In: Critical Materials Strategy for Clean Energy Technologies ISBN: 978-1-61209-829-6
Editor: Nicholas J. Bartlett

Chapter 1

CRITICAL MATERIALS STRATEGY

United States Department of Energy

FOREWORD

Each day, researchers, entrepreneurs and many others across the United States are working to develop and deploy the clean energy technologies that will enhance our security, reduce pollution and promote prosperity.

Many new and emerging clean energy technologies, such as the components of wind turbines and electric vehicles, depend on materials with unique properties. The availability of a number of these materials is at risk due to their location, vulnerability to supply disruptions and lack of suitable substitutes.

As part of the Department of Energy's efforts to advance a clean energy economy, we have developed a *Critical Materials Strategy* to examine and address this challenge.

The *Critical Materials Strategy* builds on the Department's previous work in this area and provides a foundation for future action. This Strategy is a first step toward a comprehensive response to the challenges before us. We hope it will also encourage others to engage in a dialogue about these issues and work together to achieve our Nation's clean energy goals. Ensuring reliable access to critical materials will help the United States lead in the new clean energy economy.

Steven Chu
Secretary of Energy
December 2010

ACKNOWLEDGMENTS

This report was drafted by the Office of Policy and International Affairs (PI) under the direction of Assistant Secretary David Sandalow. Principal authors were Diana Bauer, David Diamond, Jennifer Li, David Sandalow, Paul Telleen and Brent Wanner. Irving Mintzer of Mintzer Energy Group provided invaluable industry expertise, analytical insight and critical feedback. Michaelangelo Tabone (currently of University of California, Berkeley) provided extensive analytical support. Substantial contributions were made by a team of experts throughout the Department, particularly Mark Gielecki, Gwendolyn Jacobs, Tom Leckey and Fred Mayes of the Energy Information Administration; Steve Boyd, Tien Duong and Collin McCormick of the Office of Energy Efficiency and Renewable Energy; Mark Johnson of the Advanced Research Projects Agency-Energy; Linda Horton of the Office of Science; and Dave Schoeberlein and Jacqueline Zanders-Hubbard of PI. The effort benefitted from interagency input and feedback, particularly from the U.S. Geological Survey, U.S Department of Defense, White House Office of Science and Technology Policy and Council on Environmental Quality (CEQ). Other reviewers included Karl Gschneidner of Ames Laboratory and Rodd Eggert of Colorado School of Mines. The 35 respondents to a Request for Information (RFI) substantially added to our industrial insight. Navigant Consulting analyzed the RFI submissions. Energetics Incorporated provided editing services.

EXECUTIVE SUMMARY

This report examines the role of rare earth metals and other materials in the clean energy economy. It was prepared by the U.S. Department of Energy (DOE) based on data collected and research performed during 2010. Its main conclusions include:

- Several clean energy technologies—including wind turbines, electric vehicles, photovoltaic cells and fluorescent lighting—use materials at risk of supply disruptions in the short term. Those risks will generally decrease in the medium and long term.
- Clean energy technologies currently constitute about 20 percent of global consumption of critical materials. As clean energy technologies are deployed more widely in the decades ahead, their share of global consumption of critical materials will likely grow.
- Of the materials analyzed, five rare earth metals (dysprosium, neodymium, terbium, europium and yttrium), as well as indium, are assessed as most critical in the short term. For this purpose, "criticality" is a measure that combines importance to the clean energy economy and risk of supply disruption.
- Sound policies and strategic investments can reduce the risk of supply disruptions, especially in the medium and long term.
- Data with respect to many of the issues considered in this report are sparse.

In the report, DOE describes plans to (i) develop its first integrated research agenda addressing critical materials, building on three technical workshops convened by the

Department during November and December 2010; (ii) strengthen its capacity for information-gathering on this topic; and (iii) work closely with international partners, including Japan and Europe, to reduce vulnerability to supply disruptions and address critical material needs. DOE will work with other stakeholders— including interagency colleagues, Congress and the public—to shape policy tools that strengthen the United States' strategic capabilities. DOE also announces its plan to develop an updated critical materials strategy, based upon additional events and information, by the end of 2011.

DOE's strategy with respect to critical materials rests on three pillars. First, diversified global supply chains are essential. To manage supply risk, multiple sources of materials are required. This means taking steps to facilitate extraction, processing and manufacturing here in the United States, as well as encouraging other nations to expedite alternative supplies. In all cases, extraction and processing should be done in an environmentally sound manner. Second, substitutes must be developed. Research leading to material and technology substitutes will improve flexibility and help meet the material needs of the clean energy economy. Third, recycling, reuse and more efficient use could significantly lower world demand for newly extracted materials. Research into recycling processes coupled with well-designed policies will help make recycling economically viable over time.

The scope of this report is limited. It does not address the material needs of the entire economy, the entire energy sector or even all clean energy technologies. Time and resource limitations precluded a comprehensive scope. Among the topics that merit additional research are the use of rare earth metals in catalytic converters and in petroleum refining. These topics are discussed briefly in Chapter 2.

DOE welcomes comments on this report and, in particular, supplemental information that will enable the Department to refine its critical materials strategy over time. Comments and additional information can be sent to materialstrategy@energy.

The structure of this report is as follows:

Section 1 provides a brief Introduction.

Section 2 reviews the supply chains of four components used in clean energy technologies:

- *Permanent magnets* (used in wind turbines and electric vehicles)
- *Advanced batteries* (used in electric vehicles)
- *Thin-film semiconductors* (used in photovoltaic power systems)
- *Phosphors* (used in high-efficiency lighting systems)

These components were selected for two reasons. First, the deployment of the clean energy technologies that use them is projected to increase, perhaps significantly, in the short, medium and long term. Second, each uses significant quantities of rare earth metals or other key materials.

Section 3 presents historical data on supply, demand and prices. Data is provided for 14 materials, including 9 rare earth elements (yttrium, lanthanum, cerium, praseodymium, neodymium, samarium, europium, terbium and dysprosium) as well as indium, gallium, tellurium, cobalt and lithium.

Sections 4, 5 and 6 describe current programs related to critical materials within DOE, the rest of the federal government and other nations.

Section 7 presents supply and demand projections. Potential supply/demand mismatches are identified and shown graphically. Complexities that complicate market response to these mismatches are also discussed.

Section 8 presents "criticality assessments"— analyses that combine the importance of a material to the clean energy economy and supply risk with respect to that material. The analytical approach is adapted from a methodology developed by the National Academy of Sciences (NAS 2008). The analyses may be useful in priority-setting for research and other purposes. Applying this methodology to the materials listed above, terbium, neodymium, dysprosium, yttrium, europium and indium have greatest short-term "criticality" (Figure ES-1). All of these materials except indium remain critical in the medium term (Figure ES-2).

Chapter 9 discusses program and policy directions. Eight broad categories are considered: (i) research and development, (ii) information gathering, (iii) permitting for domestic production, (iv) financial assistance for domestic production and processing, (v) stockpiling, (vi) recycling, (vii) education and (viii) diplomacy. These programs and policies address risks, constraints and opportunities across the supply chain, as shown in Figure ES-3. DOE's authorities and historic capabilities with respect to these categories vary widely. Some (such as research and development) relate to core competencies of DOE. Others (such as permitting for domestic production) concern topics on which DOE has no jurisdiction. With respect to research and development, topics identified for priority attention include rare earth substitutes in magnets, batteries, photovoltaics and lighting; environmentally sound mining and materials processing; and recycling. The chapter ends with a summary of recommendations.

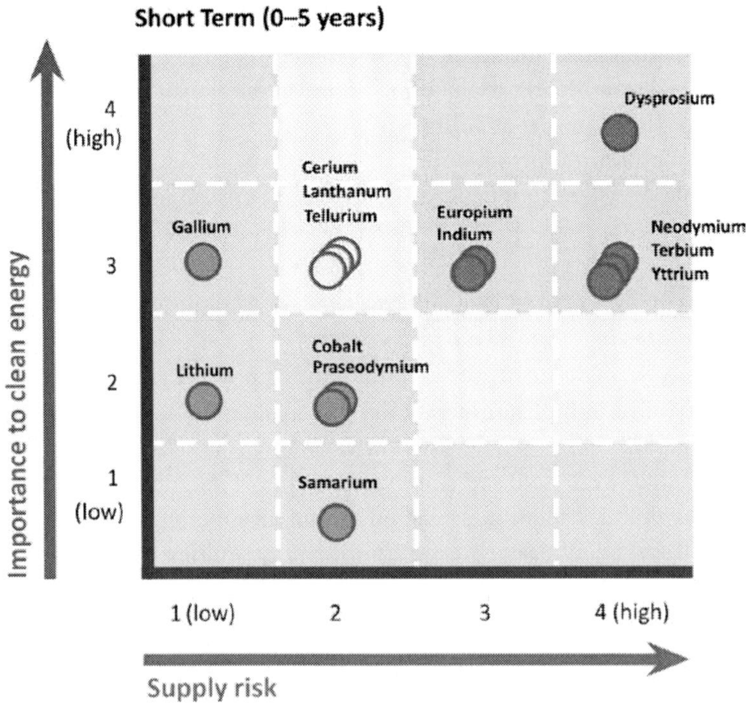

Figure ES-1. Short-term criticality matrix.

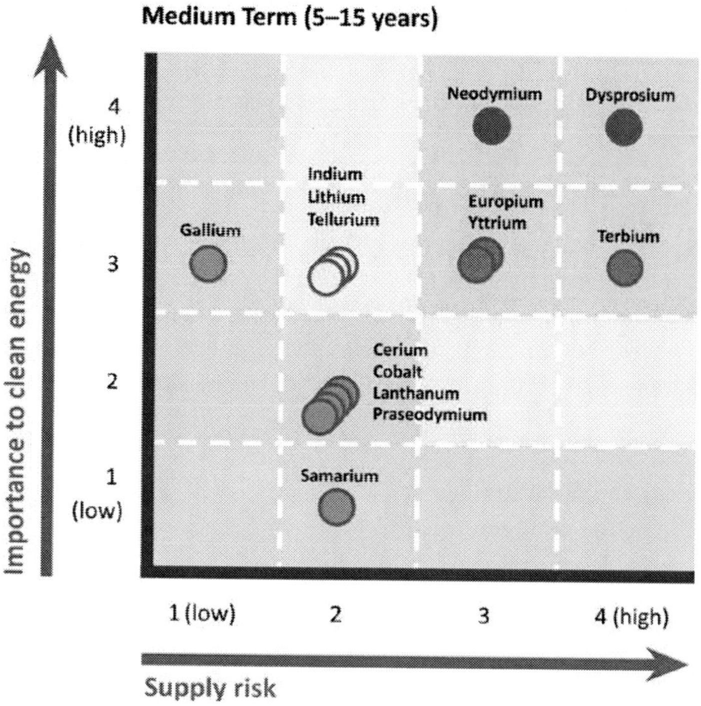

Figure ES-2. Medium-term criticality matrix.

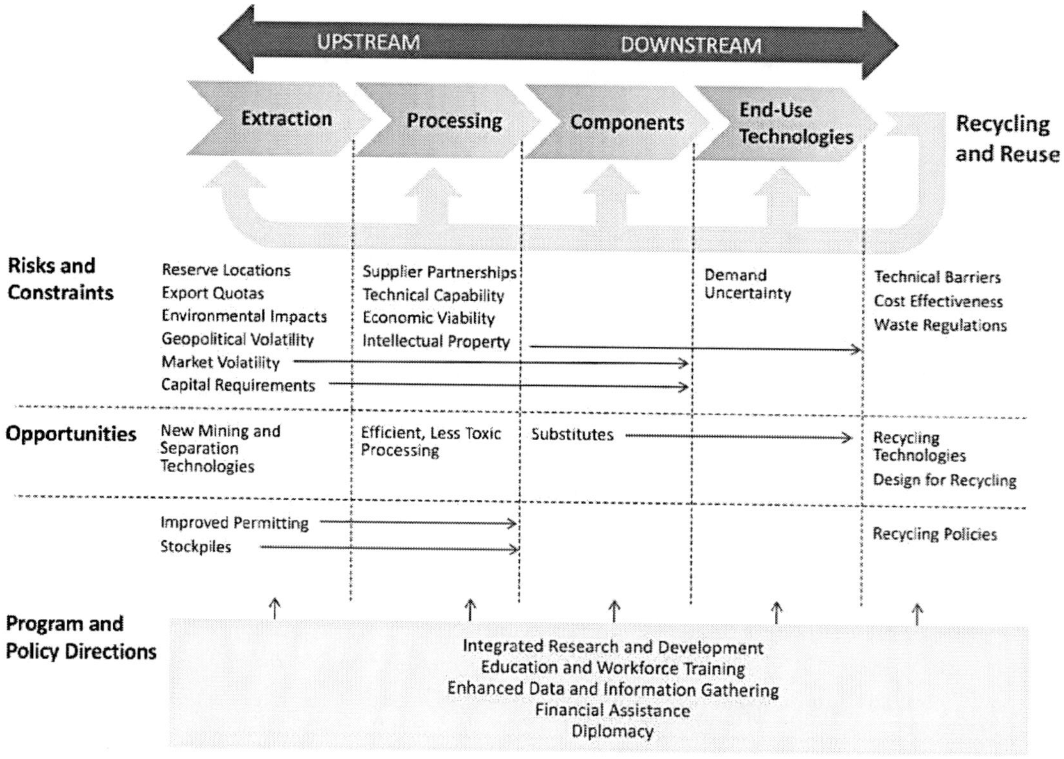

Figure ES-3. Program and policy directions and the critical material supply chain.

1. Introduction

This report examines the role of rare earth metals and other materials in the clean energy economy. The report focuses in particular on the role of key materials in renewable energy and energy-efficient technologies. Deployment of these technologies is expected to grow substantially in the years ahead. Yet many of these technologies—including wind turbines, electric vehicles, solar cells and energy-efficient lighting—depend on components often manufactured with these materials.

The U.S. Department of Energy (DOE) has worked on topics related to materials for many years. However, before 2010, that work was not coordinated across different DOE offices and programs. Accordingly, DOE has developed this report for the following purposes:

- Assess risks and opportunities
- Inform the public dialogue
- Identify possible program and policy directions

1.1. Scope of the Report

This report addresses both the short- (0–5 years) and medium-term (5–15 years) deployment of wind turbines, electric vehicles, solar cells and energy-efficient lighting. These technologies were selected for two reasons. First, they are expected to be deployed substantially within the global clean energy economy over the next 15 years. Second, they use less common materials and could, through their deployment, substantially increase global demand for these materials.

The report focuses on a small number of illustrative scenarios, rather than developing an exhaustive set. Reference and policy-based scenarios are used to develop low and high plausible estimates for materials consumption over the short and medium term. International scenarios and roadmaps are used, with some attention to the U.S. dimension.

The scope of this report is limited. It does not address the material needs of the entire economy, the entire energy sector or even all clean energy technologies. Time and resource limitations precluded a comprehensive scope. Among the topics that merit additional research are the use of rare earth metals in catalytic converters and in petroleum refining. These topics are discussed briefly in Chapter 2.

1.2. Materials Analyzed

Fourteen elements and related materials were selected for a criticality assessment within this report (Figure 1-1). Eight of these are rare earth metals, which are valued for their unique magnetic, optical and catalyst properties. The materials are used in clean energy technologies as follows. Lanthanum, cerium, praseodymium, neodymium, cobalt and lithium are used in electric vehicle batteries. Neodymium, praseodymium and dysprosium are used in magnets for electric vehicles and wind turbines. Samarium is also used in magnets. Lanthanum, cerium, europium, terbium and yttrium are used in phosphors for energy-efficient lighting. Indium, gallium and tellurium are used in solar cells. The materials were selected for study based on factors contributing to risk of supply disruption, including a small global market, lack of supply diversity, market complexities caused by coproduction and geopolitical risks.

While these materials are generally used in low volumes relative to other resources, the anticipated deployment of clean energy technologies will substantially increase worldwide demand. In some cases, clean energy technology demand could compete with a rising demand for these materials from other technology sectors.

Not all of the materials examined in the report are critical. Until a criticality assessment is presented in Chapter 8, the materials of interest examined in the report will be referred to as "key materials."

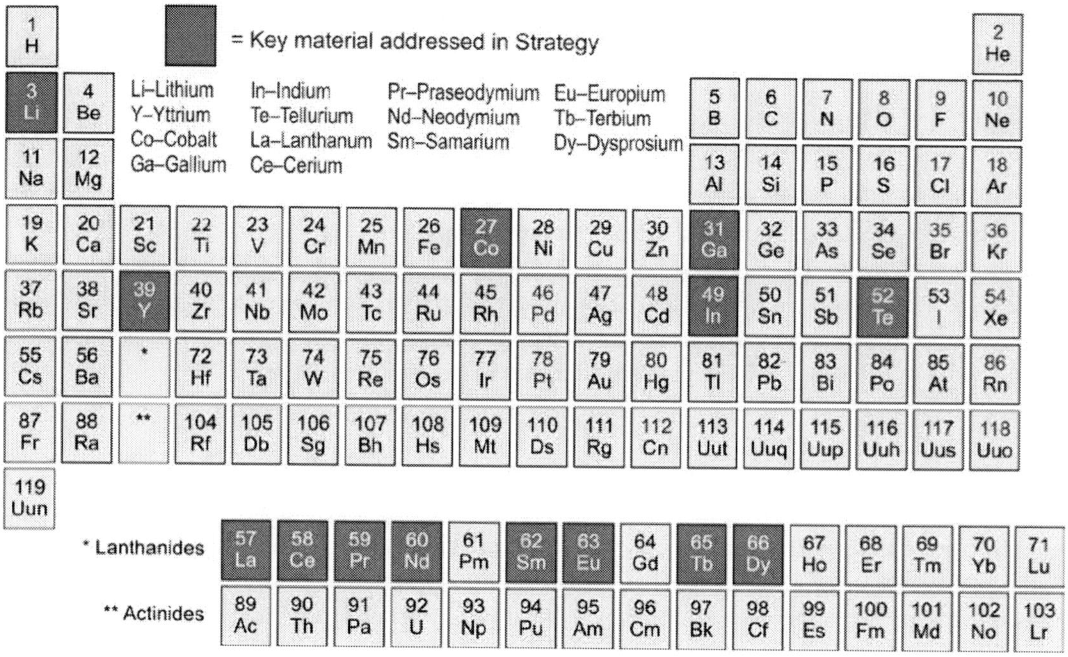

Figure 1-1. Key materials within the periodic table of the elements.

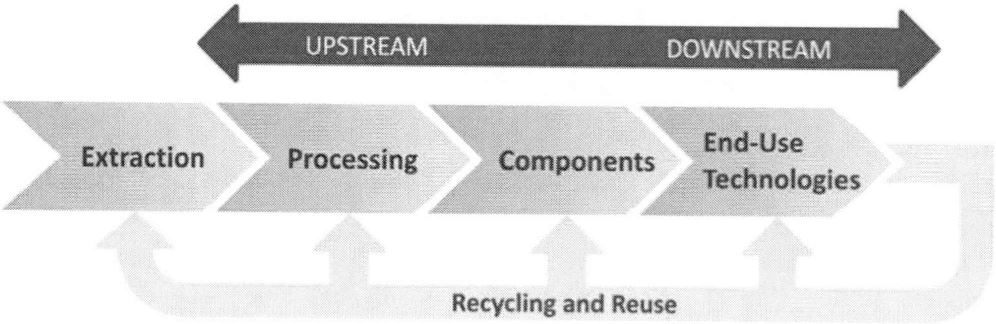

Figure 1-2. Basic materials supply chain.

1.3. Supply Chain Framework

Maintaining the availability of materials for clean energy is not simply a mining issue. Manufacturing processes across the full supply chain must also be considered.

The industrial supply chain in Figure 1-2 illustrates the steps by which materials are extracted from mines, processed and transformed into useful components or utilized in end-use applications. The specific industrial supply chain for each material and component varies, but in general it can be described by the generic supply chain. The supply chain provides a useful context in which to explore the technical, geopolitical, economic, environmental and intellectual property factors that impact the supply of these materials and the technologies that use them. In addition, a supply chain framework can inform where to target potential policy tools.

Elemental materials are extracted from the earth via mining. Next, they are processed via separation and refining to obtain the desired composition or purity. Materials may be extracted either as major products, where ore is directly processed to extract the key materials or they may be coproducts or byproducts of other mining operations. The coproduction and byproduction processes create complex relationships between the availability and extraction costs of different materials, which may cause supply curves and market prices to vary in ways not captured by simple supply and demand relationships.

Processed materials are used to manufacture component parts that are ultimately assembled into end-use technologies. The generic supply chain also shows the potential for recycling and reusing materials from finished applications, though materials can be reclaimed at any stage of the supply chain and reused either upstream or downstream.

1.4. Formulating a Strategy

Current global materials markets pose several challenges to the growing clean energy economy. Lead times with respect to new mining operations are long (from 2–10 years). Thus, the supply response to scarcity may be slow, limiting production of technologies that depend on such mining operations or causing sharp price increases. In addition, production of some materials is at present heavily concentrated in one or a small number of countries. (More than 95% of current production capacity for rare earth metals is currently in China.) Concentration of production in any supplier creates risks for global markets and creates geopolitical dynamics with the potential to affect other strategic interests of the United States. Value-added processing and some patent rights are also concentrated in just a few countries, creating similar risks. For some materials, these factors are likely to lead to future material supply-demand mismatches. In many instances, the private sector market will likely respond to correct the imbalance. This report offers a number of policy and program directions to supplement market response where this is warranted.

The approach to proactively address material supply risks and prevent supply chain disruptions, while building a robust clean energy economy, has three elements:

- Achieve globally diverse supplies
- Identify appropriate substitutes
- Improve our capacity for recycling, reuse and more efficient use of critical materials

This is the first in a series of critical materials strategies. DOE expects to update the Strategy regularly to reflect feedback received and changing circumstances and intends to issue an updated *Critical Materials Strategy* next year.

2. USE OF KEY MATERIALS IN CLEAN ENERGY TECHNOLOGIES

2.1. Introduction

This chapter describes the use of key materials in the components of several clean energy technologies. The chapter focuses on the following:

- **Permanent magnets** used in wind turbines and electric drive vehicles
- **Batteries** used in vehicles with electric drive trains
- **Thin films** used in photovoltaic (PV) cells
- **Phosphors** used in fluorescent lighting

These components and technologies were selected for priority attention because (i) each currently relies on critical materials, including rare earth metals and (ii) demand for each may grow significantly in the short and medium term.

Table 2-1 provides an overview of the key materials used in leading clean energy technologies.

Table 2-1. Materials in Clean Energy Technologies and Components.

		CLEAN ENERGY TECHNOLOGIES AND *COMPONENTS*				
		Solar Cells	Wind Turbines	Vehicles		Lighting
	MATERIAL	*PV films*	*Magnets*	*Magnets*	*Batteries*	*Phosphors*
Rare Earth Elements	Lanthanum				●	●
	Cerium				●	●
	Praseodymium		●	●	●	
	Neodymium		●	●	●	
	Samarium		●	●		
	Europium					●
	Terbium					●
	Dysprosium		●	●		
	Yttrium					●
	Indium	●				
	Gallium	●				
	Tellurium	●				
	Cobalt				●	
	Lithium				●	

This chapter describes the four components (magnets, batteries, PV thin films and phosphors) in more detail, with an emphasis on factors that influence the amount of key materials required for each. These factors include product design choices (such as battery chemistry, motor specification, phosphor composition and PV film thickness), technical innovations to reduce the amount of key material within a product required for a given performance level and measures to reduce manufacturing processing losses.

This chapter also provides examples of the supply chain by which key materials are mined, processed and ultimately incorporated into clean energy products by manufacturers. Insights from the supply chain will help determine which materials face supply risks in the future and identify opportunities to mitigate those risks.

2.2. Permanent Magnets

Permanent magnets (PMs) produce a stable magnetic field without the need for an external power source and are a key component of lightweight, high-power motors and generators. PM generators are used in wind turbines to convert wind energy into electricity, while PM motors are used in electric vehicles (EVs), hybrid-electric vehicles (HEVs) and plug-in hybrid-electric vehicles (PHEVs) to convert energy stored in the vehicle's battery into mechanical power for propulsion.

Material Content

The use of certain rare earth elements (REEs) in PMs significantly reduces the weight of the motor or generator for a given power rating. Current hybrid-electric drive designs employ motors with neodymium-iron-boron (NdFeB) magnets.[1] Large capacity wind turbines (with several megawatts [MW] or more of power generation capacity) increasingly use rare earth PM generators. Although these turbines still represent a relatively small portion of the wind market, their share is likely to grow as purchasers increasingly choose larger turbines for wind projects.

The rare earth content of NdFeB magnets varies by manufacturer and application. An electric drive vehicle may use up to a kilogram (kg) of Nd, while each wind turbine may contain several hundred kilograms. Rare earth PMs may also incorporate praseodymium (Pr), which can be substituted for or combined with Nd. Dysprosium (Dy) or terbium (Tb) may also be added to the intermetallic alloy to increase the temperature at which the magnet can operate before losing its magnetic field (London 2010). Specific material intensity estimates for vehicle motors and wind turbine are discussed in Chapter 7 and Appendix B. The cumulative demand for Nd and other REEs in these clean energy technologies is a function of both the material content per individual product and the total number of products sold. Therefore, aggressive technology penetration rates envisioned under many worldwide clean energy strategies could significantly increase global demand for Nd, Pr, Dy and Tb.

Supply Chain Example: Rare Earth Permanent Magnets

The PM supply chain begins with the extraction and separation of Nd and other REEs from ore. Depending on the geographic location of the mine and composition of the ore, REE coproduction may be a byproduct of extraction of another ore, such as iron ore containing a mixture of REEs in varying concentrations. The vast majority of REE mining currently occurs in China.

Once mined or coproduced, REE ore can be separated into a concentrate, processed into a mixed rare earth solution and elementally separated to oxides by solvent extraction. Rare earth oxides are ultimately used to produce rare earth metals, alloys and powders, which manufacturers use as the building blocks for components of clean energy technologies.

Intellectual property plays a significant role in the supply chain. Manufacturers employ proprietary variations of elements within the magnets to produce the desired properties and proprietary process technologies for forming magnetic shapes via bonding or sintering. Sintering produces higher performance magnets required for electric drive and larger wind turbine applications, while bonded magnets are sufficient for use in other applications, such as electronics.

Master patents on NdFeB magnets are controlled by two firms: Hitachi Metals (formerly Sumitomo) in Japan and Magnequench, a former U.S. firm that was sold to a Chinese-backed consortium in 1995 (Dent 2009).[2] Hitachi has used this intellectual property protection to capture a large portion of the market for high-quality magnetic materials, while the Magnequench sale gave Chinese companies access to the intellectual property and technology necessary to establish production plants and further increase supply chain integration. Licensed production of sintered NdFeB magnets is currently limited to 10 firms in China, Japan and Germany. Relevant patents for sintered NdFeB magnets may expire in 2014 (Arnold Magnetic Technologies 2010).

Figure 2-2 illustrates the supply chain for vehicle and wind turbine applications using NdFeB PMs. The reuse and recycling loop in the generic material supply chain (Figure 1-2) is not shown in Figure 2-2 because there is currently only limited recovery of manufacturing waste and no measurable recovery from aftermarket products (Arnold Magnetic Technologies 2010). However, improved designs for recycling coupled with larger streams of materials could eventually allow for the economical recycling and reuse of magnetic materials.

2.3. Batteries

Batteries are a key component in vehicle applications—HEVs, PHEVs and EVs all require batteries to store energy for vehicle propulsion. HEVs rely on an internal combustion engine as the primary power source, but use a battery to store excess energy captured during vehicle braking or produced by the engine. The stored energy provides power to an electric motor that can assist in acceleration or provide limited periods of primary propulsion power. PHEV configurations vary, but generally incorporate a higher-capacity battery than HEVs, which can be recharged externally and used as the primary power source for longer durations and at higher speeds than is required for a HEV. In EVs, the battery is the sole power source.

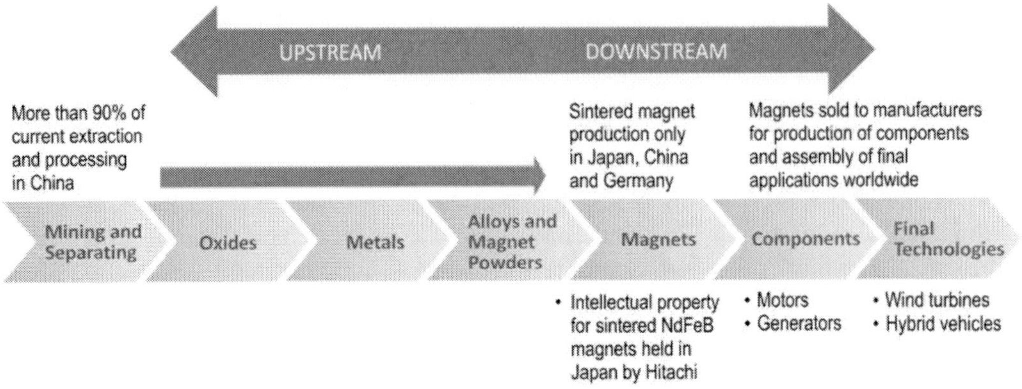

Figure 2-1. Supply chain for rare earth element PM applications.

Material Content

The electric drive vehicles described above all require rechargeable (also called "secondary") batteries with the capacity to rapidly store and release electrical energy over multiple cycles. There are a wide variety of battery chemistries available. Current generation HEVs use nickel metal hydride (NiMH) batteries. The most common NiMH chemistries use a cathode material designated as AB_5. A is typically rare earth mischmetal containing lanthanum, cerium, neodymium and praseodymium; while B is a combination of nickel, cobalt, manganese and/or aluminum (Kopera 2004). A current- generation hybrid vehicle battery may contain several kg of REE material. Specific material intensity estimates are discussed in Chapter 7 and Appendix B.

PHEVs and EVs require greater storage capacity and higher power ratings than HEVs and consequently are likely to employ lithium-ion batteries (National Research Council 2010). Although battery manufacturers are still working to address cost and safety issues, lithium-ion chemistries offer better energy density, cold-weather performance, abuse tolerance and recharge rates than NiMH batteries (Vehicle Technologies Program 2009). Thus, the demand for lithium and other materials associated with lithium-ion battery chemistries will likely grow substantially with the wide- scale deployment of EVs and PHEVs. Lithium-ion batteries that show promise for electric vehicle applications typically use either graphite as the anode and some form of lithium salt in both the cathode and electrolyte solution. The lithium content per vehicle battery varies widely depending on manufacturer design choices. Researchers from Argonne National Laboratory have estimated that a battery capable of providing 100 miles of range for an electric vehicle would contain between 3.4 and 12.7 kg of lithium, depending on the specific lithium-ion chemistry used and the battery range (Gains and Nelson 2010).

Supply Chain Example: Lithium-Ion Batteries

Lithium for battery cathode and electrolyte materials is produced from lithium carbonate, which is most widely and economically extracted from salt lake brine deposits via a lime soda evaporation process. The process starts with concentrating the lithium chloride by evaporating salty water in shallow pools for 12 to 18 months, which is then treated with sodium carbonate (soda ash) to precipitate out the lithium carbonate. Lithium carbonate has also been produced from spodumene (a silicate of lithium and aluminum) and hectorite clay deposits, but recovery from these sources is more expensive (USGS 2009a). Currently, Chile is the largest producer of this lithium carbonate.

Manufacturers produce battery cells from anode, cathode and electrolyte materials. All Li-ion batteries use some form of lithium in the cathode and electrolyte materials, while anodes are generally graphite based and contain no lithium.[3] These cells are connected in series inside a battery housing to form a complete battery pack. Despite lithium's importance for batteries, it represents a relatively small fraction of the cost of both the battery cell and the final battery cost (Deutsche Bank 2009).

While some lithium-ion batteries are available in standard forms and sizes, most are designed to meet the requirements of a specific product (Brodd 2005). This design process entails close coordination between battery manufacturers and automakers to develop batteries with suitable performance for electric drive vehicles.

Various programs seek to recover and recycle lithium-ion batteries. These include prominently placed recycling drop-off locations in retail establishments for consumer electronics batteries, as well as recent efforts to promote recycling of EV and PHEV batteries as these vehicles enter the market in larger numbers (Hamilton 2009). Current recycling programs focus more on preventing improper disposal of hazardous battery materials and recovering battery materials that are more valuable than lithium. However, if lithium recovery becomes more cost effective, recycling programs and design features provide a mechanism to enable larger scale lithium recycling. Another potential application for lithium batteries that have reached the end of their useful life for vehicle applications is in stationary applications such as grid storage.

The supply chain for the production of lithium for use in lithium-ion batteries, shown in Figure 2-2, is illustrative of the supply chain for many types of batteries. It involves multiple, geographically distributed steps and it overlaps with the production supply chains of other potential critical materials, such as cobalt, which are also used in battery production.

2.4. Polycrystalline Photovoltaic Thin Films

Photovoltaic (PV) technologies are the most common solar technologies used to generate electricity. PV technologies include crystalline silicon, thin films, high-efficiency III-V cells with optical concentrators and nanotechnology-based films. In 2008, conventional crystalline silicon-based cells were the dominant PV technology, accounting for 86% of the total global PV market (USGS 2009b). However, thin film technologies constituted a growing share of the remainder of the market.[4] They are increasingly prominent among PV technologies due to several advantages relative to traditional crystalline silicon "thick films": they require less functional material; they can be manufactured in continuous rolls or sheets; and they can be deposited on flexible substrates.

Material Content

Three primary material formulations for PV thin films are on the market: amorphous silicon (a-Si), copper-indium diselenide (CIS) and cadmium telluride (CdTe). While a-Si accounted for about 50% of the thin-film market in 2008, the shares of both CIS and CdTe are increasing. A major subset of CIS thin films are copper-indium gallium diselenide (CIGS) formulations, which are about 10% copper, 28% indium, 10% gallium and 52% selenium. About 2 tonnes[5] of copper, 4 tonnes of indium and 2 tonnes of gallium were purchased to produce 158 MW of CIGS solar cells in 2008. About 100 tonnes of tellurium was purchased in 2008 to produce cadmium tellurium material for 358 MW of CdTe cells (USGS 2009b). This total of about 500 MW in CIS and CdTe accounted for roughly 5% of the global PV market in 2008. Other thin film technologies are being developed, such as copper zinc tin sulfide.

Next-generation technologies include organic solar cells and dye-sensitized solar cells (EERE 2010). Some researchers recommend focusing on lower cost, more readily available materials. Potential composite material formulations include iron sulfide (FeS_2), copper sulfide (Cu_2S) and zinc phosphide (Zn_3P_2) (Wadia et al. 2009, 2072–2077), although these are still in the very early stages of development.

Supply Chain Example: PV Thin Films

Tellurium, gallium and indium all have fairly diverse sources of global production, though material production for PV films can depend on coproduction with non-ferrous metals. For example, tellurium is produced as a secondary product from copper refining. After mining or secondary production, materials are purified, either as individual elements or as compounds, such as CdTe. For CdTe, this step occurs in Canada (USGS 2009b).

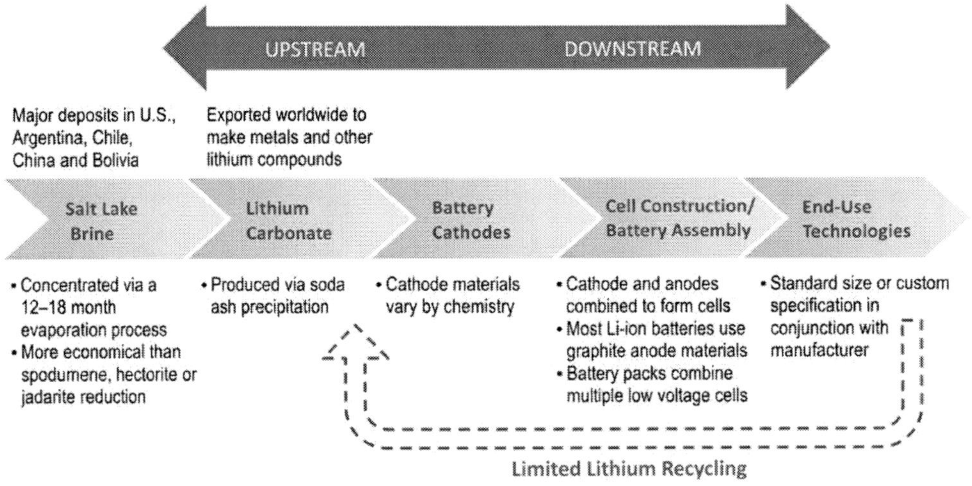

Figure 2-2. Supply chain for lithium-ion batteries.

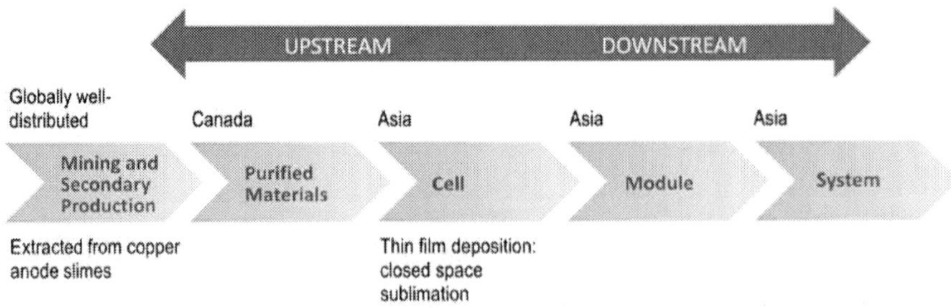

Figure 2-3. CdTe supply chain.

The purified materials are deposited in multiple layers onto transparent conducting oxide substrates to form PV cells. A current typical deposition thickness for a CdTe film is 2 microns[6]. This step is analogous to semiconductor manufacturing. Several different deposition techniques can be used, ranging from sputtering or coevaporating for CIGS, closed-space sublimation for CdTe and solution- based printing approaches for more early-state technologies (Rose 2009). Finally, cells are made into modules, which are then made into systems.

Consistent with semiconductor manufacturing, these final steps often take place in Asia due to lower labor costs and local policies favorable to manufacturing. Some recent moves toward increased U.S. production have occurred, due in part to American Recovery and Reinvestment Act investments. These include the DOE Loan Guarantee and Manufacturing Tax Credit programs. In general, there is intense focus on reducing costs while maintaining a diversity of manufacturing technology options as the industry tries to reduce its production costs per system watt (Rose 2009). Much of the R&D focus is on cell production, including reducing the quantities of required specialty metals such as tellurium, indium and gallium. Figure 2-3 illustrates the manufacturing supply chain for CdTe thin films.

COMPARISON OF LIGHTING TECHNOLOGIES

Incandescent lamps generate visible light by heating a filament in a vacuum or inert gas to produce light. They are simple and inexpensive to manufacture, but relatively inefficient.

Fluorescent lamps generate visible light by using electricity to excite mercury vapor inside a tube, causing it to emit ultraviolet light, which excites a phosphor compound coating the inside of the tube. Fluorescent lamps are more complicated and costly to manufacture than incandescent, but have lifbecause they a more cost effectivee cycle are more energy efficient and last longer. **Linear fluorescent lamps (LFL)** are common in industrial and commercial buildings and newer **compact fluorescent lamps (CFL)** are designed to fit existing lamp sockets in residential buildings.

Light Emitting Diodes (LEDs) produce visible light using the electroluminescence of a compound semiconductor crystalline material. This process is potentially more energy efficient than either incandescence or fluorescence. When connected to a flow power source, the of current triggers a quantum mechanical process inside the diode, which produces light in specific colors (usually redgreen or blue). White light is created by combining the light from these colored LEDs or by coating a blue LED with yellow phosphor (Department of Energy 2008).

Organic LEDs (OLEDs) produce visible light when an electrical charge is applied to extremely thin organic materials layered between two electrodes. The technology is still in the early stages of development, but has the potential to efficiently produce visually appealing white lighting in a thin, flexible form that could compete directly with fluorescent and conventional LED lighting.

2.5. Phosphors and Lighting

Improvements in lighting efficiency provide opportunities to significantly reduce energy demand. Lighting accounts for approximately 18% of electricity use in U.S. buildings—second only to space heating (DOE 2009). Lighting technologies can be broadly grouped into four categories: traditional incandescent, fluorescent, light emitting diodes (LEDs) and organic light emitting diodes (OLEDs).[7] Many of these lighting technologies incorporate key materials, including REEs. This Strategy is primarily concerned with REEs used in fluorescent lighting phosphors.

Material Content

Though older fluorescent lighting designs are REE-free, the current generation of more efficient, spectrally complete and visually pleasing lamps uses phosphors containing different concentrations of lanthanum, cerium, europium, terbium and yttrium to achieve various lighting effects. Although exact cost breakdowns for light bulbs are proprietary, phosphors are thought to represent a significant portion of the cost of an LFL or CFL. Phosphors accounted for 7% of all REE usage by volume and 32% of the total value in 2008 (Kingsnorth 2010). The exact composition of phosphors, including REE variety and weight percentages, differs by manufacturer and is considered proprietary information.

Emerging lighting technologies have dramatically lower REE content than fluorescent lamps. White LED designs eliminate the need for lanthanum and terbium phosphors, but may still use cerium and europium phosphors to convert blue LEDs to useful white light. Gallium and indium are used in the formation of the LED compound semiconductor material. Some manufacturers add neodymium as a glass component to shift the color of certain products to more closely resemble natural light. However, in 2010 this use represented a very small percentage of overall neodymium use (General Electric 2010). OLEDs can be free of all lanthanides, but bulb manufacturers may still use other key materials such as indium. Figure 2-4 highlights the differences in material content between competing lighting technologies.

In the short and medium terms, the demand for LFL and CFL fluorescent lamps using REEs in their phosphor formulations is expected to increase. LFLs will continue to dominate the commercial lighting market while CFLs will increasingly displace incandescent products in the residential lighting sector. DOE standards for general service fluorescent lamps, issued in June 2009, mandate increased efficiency (lumens per watt) ratings for different classes of fluorescent lighting, effectively phasing out most non-REE LFL light bulbs. Additional regulations require progressively higher efficiency standards for incandescent light bulbs starting in 2012, effectively phasing them out in favor of CFL light bulbs. The National Electrical Manufacturers Association, a lighting industry trade group, estimates that the new rules for general service fluorescent lamps will increase the demand for REE phosphors by 230% over current levels (NEMA 2010), though some of this increase in demand reflects a market shift that is already under way.

In the long term, LED and OLED technologies will likely capture a significant share of the lighting market as their cost and performance make them increasingly competitive with fluorescent technologies. This change could mitigate the demand increase for REE phosphors.

Supply Chain Example: Phosphors in Fluorescent Lighting

REEs used in phosphors must be 99.999% pure, necessitating tight control over the manufacturing process.[8] The presence of impurities of a few parts per million can distort the color characteristics of a given phosphor. In order to achieve these high purities, the purification takes many more separation stages, significantly increasing the cost of the rare earth oxides (REOs) used to produce the phosphors. Suppliers of phosphors used in lighting products generally produce mass quantities of similar phosphor materials for application in television screens, computer monitors and electronic instrumentation (McClear 2008).

China currently consumes 80% of world's lighting phosphor supply to produce components for major lighting manufacturers, although it subsequently exports the majority of these components for sale worldwide. The location of the lamp manufacturing process (which includes the production of glass tubes, coating with phosphors and assembly of bulb components) is driven by the labor and transportation costs of different types of bulbs, as well as by local government manufacturing incentives.

CFLs are manufactured almost exclusively in China and distributed by major lighting manufacturers for sale worldwide. LFLs are still primarily assembled in plants in North America and Europe that are closer to the ultimate points of sale. This arrangement exists because it is much cheaper to ship the raw materials than the LFL bulbs, whose volume consists mostly of air inside the fragile lighting tubes.

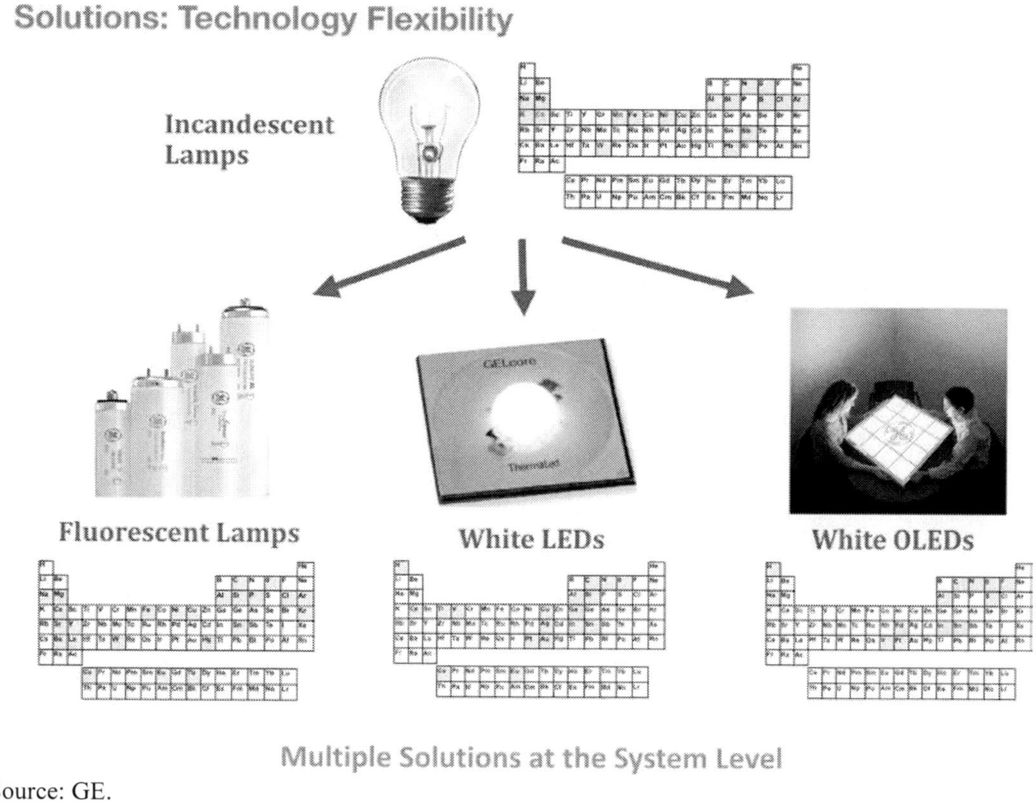

Source: GE.

Figure 2-4. Material usage comparison for lighting technologies.

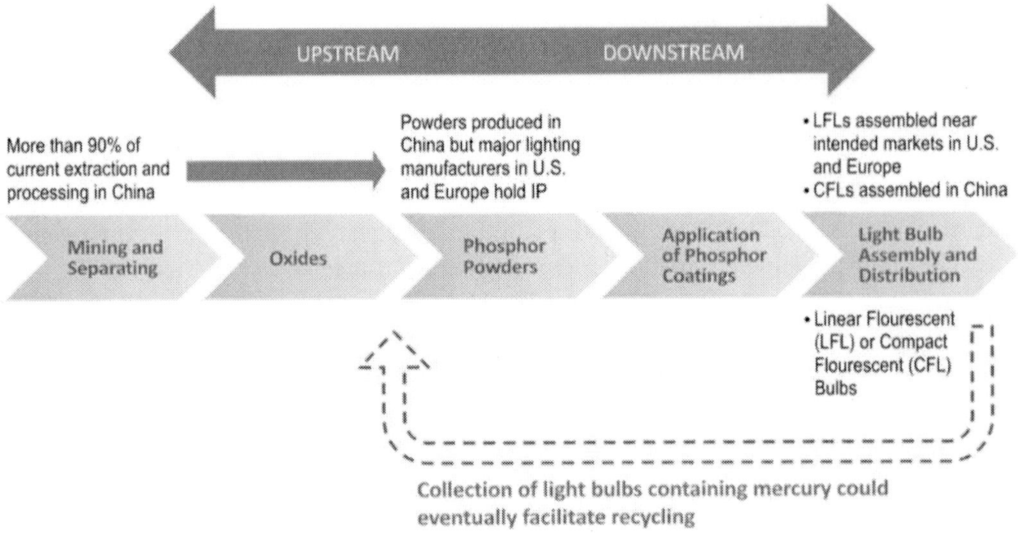

Figure 2-5. Supply chain for rare earth phosphors in fluorescent lighting.

Regardless of manufacturing and assembly location, major U.S. lighting manufacturers continue to hold the intellectual property rights to formulas for the fluorescent lighting phosphors and invest significantly in research and development (R&D) related to lighting manufacturing. This allows U.S. firms to retain control of the value chain, despite the large role of Chinese firms in the manufacturing process.

Phosphors and component REEs are not currently recovered from fluorescent bulbs, but due to the mercury content, there is a growing infrastructure to recover used LFLs and CFLs for safe disposal.

This infrastructure could eventually facilitate the recycling of REEs from bulbs. A simplified supply chain for fluorescent lighting is shown in Figure 2-5.

Other Technologies

The technologies and clean energy applications discussed in this chapter were selected because they are the most likely to see wide-scale commercialization and deployment in the short to medium term. Therefore, they are most likely to drive clean energy demand for the key materials of interest in the short and medium term. However, there are a number of other clean energy applications using key materials that are outside the scope of this version of the Strategy, but are worth noting:

Grid Storage Batteries play an essential role in clean energy generation and distribution by storing energy that is generated in excess of current demand for later use. This grid storage capability is particularly important for wind and solar power electricity generation where generation capacity fluctuates with the available wind or light. These applications could therefore employ other types of battery technologies that are more easily scaled up in size for the large capacity storage requirements than lithium ion or NiMH. The Office of Electricity Delivery and Energy Reliability is investing in several large-scale battery-based grid storage demonstrations, including lithium-ion, sodium-sulfur, lead-carbon and iron-chromium technologies. Even older rechargeable battery technologies, such as lead acid or nickel cadmium, may be suitable and cost advantageous, since the storage batteries do not need to be as lightweight or compact as those used in vehicle applications.

Fuel Cells are a promising clean energy technology for vehicle propulsion and distributed power generation. REEs are used in several different fuel cell chemistries. In particular, there is no substitute for their use in solid oxide fuel cell separator stacks. However, fuel cell vehicles are unlikely to see large-scale commercialization in the short to medium term, due to both technical challenges in cost-effective fuel cell designs and the lack of a hydrogen refueling infrastructure.

Nuclear Power technologies incorporate some of the key materials considered in this Strategy and are often classified as a clean energy technology. However, the high capital costs and lengthy permitting requirements for new nuclear power plants make it unlikely that nuclear power's share of key material usage will expand rapidly in the short to medium term.

Electric Bicycles use NdFeB permanent magnet motors and batteries in a manner similar to EVs and PHEVs. The motor and battery size and key material content per bicycle is very small compared to electric drive automobiles, but electric bicycle sales are sold in much greater numbers—particularly in developing countries. Electric bicycles are not included in the clean energy demand projections in Chapter 7 of this Strategy, but they could still represent a significant share of the growth in demand for Nd and other REEs used in PMs.

Magnetic Refrigeration shows great promise for improving the energy efficiency of the refrigeration process using rare earth PMs. Some experts believe this technology could be commercialized and capture a significant share of the refrigeration market in the medium term. However, this technology was not considered in the clean energy material demand projections in Chapter 7 due to uncertainties about the timeline for commercialization, projected demand and material intensity of the commercial products.

Additionally, several other energy-related technologies use significant quantities of REEs:

Fluid Cracking Catalysts (FCCs) are used in the oil refining process to convert heavy oils (gas oils and residual oils) into more valuable gasoline, distillates and lighter products. Rare earth elements are used in FCC catalysts to help control the product selectivity of the catalyst and produce higher yields of more valuable products such as gasoline. Lanthanum is the predominant REE used in FCCs, along with lesser amounts of cerium and neodymium. Cerium is also a key component of FCC additives that are used to help reduce stationary source nitrogen oxide (NO_x) and sulfur oxide (SO_x) pollutants. According to personal communications with a catalysts supplier, the estimated world demand in 2009 for REOs used in FCC catalysts was approximately 7,550 tonnes/year. DOE has estimated that the U.S. refinery industry consumption of REOs for FCC catalysts is approximately 3,500 to 4,000 tonnes/year[9], REEs used in FCCs represent a very small fraction of the overall cost of gasoline and other petroleum products and are not required for refining. However, a disruption in REE supply could have a noticeable impact on refinery yields and require capital investments to re-optimize the fluid cracking process for operation without REEs.

Automotive catalytic converters use cerium to facilitate the oxidation of carbon monoxide (CO), helping significantly reduce vehicle CO emissions. While the amount of cerium required per vehicle is very small, catalytic converters are used in practically every passenger vehicle and accounted for approximately 9% of total U.S. rare earth use consumption in 2008.

These technologies may be considered for further analysis in future revisions to this report.

3. Historical Supply, Demand and Prices for the Key Materials

This chapter presents historical data on supply, demand and prices. Data is provided for fourteen materials, including nine rare earth elements (yttrium, lanthanum, cerium, praseodymium, neodymium, samarium, europium, terbium and dysprosium) as well as indium, gallium, tellurium, cobalt and lithium.

Text Box 3-1

Resources: A concentration of naturally occurring materials in such form that economic extraction of a commodity is regarded as feasible, either currently or at some future time.

Reserves: Resources that could be economically extracted or produced at the time of determination. The term reserves need not signify that extraction facilities are in place and operative.

Source: U.S. Geological Survey http://minerals.usgs.gov/minerals/pubs/mcs/2010/mcsapp2010.pdf

3.1. Supply

The supply of a material is a function of resources, reserves and production. "Resources" include identified and undiscovered resources. Within identified resources there is further differentiation between demonstrated and inferred resources. For the short- to medium- term focus of this report, we consider demonstrated resources only. "Reserves" refer to resources that can be extracted economically at the time of determination, but may extend beyond the medium term if new infrastructure is necessary before bringing the mine online.

Production generally occurs in countries with large resources and reserves, but exceptions exist. In some cases, small reserve holders may also produce the material, while countries with no reserves could be a major refinery producer of imported primary or raw material.

Table 3.1 reviews the production characteristics of key materials, the top ranked countries for mining and refining, U.S. production (if applicable) and top reserve holding countries for rare earths.

Reserves and Production of Rare Earth Elements

Rare earth metals are widely distributed across the earth. China holds around 36% of the REE reserves, Russia and other members of the Common wealth of Independent States (CIS) hold 19%, the U.S. holds around 13% and Australia has 5%. REEs occur in dilute concentrations in ores of other minerals. The light rare earths (atomic numbers 57–61), such as lanthanum and neodymium and medium rare earths (atomic numbers 62–64), such as europium, are found mainly in bastnäsite and monazite.[19] Heavy rare earths (atomic numbers 65–71), such as terbium and dysprosium, along with yttrium (atomic number 39), are somewhat more scarce and often concentrated in ionic adsorption clay and xenotime, commonly found in southeastern China (USGS 2010j).

Table 3-1. Production and Reserves Information on Key Materials[10]

	Production characteristics	2009 top-ranked global primary and refinery producers plus U.S.-related information (in *tonnes* unless otherwise indicated)				Top-ranked reserve holding countries, in rank order		Total global reserves (in *tonnes)*
		Mine production:		*Refined metal:*				
Rare earth elements (in rare earth oxide/ REO)	Occur in dilute concentrations in metal ores. Often co-produced with other metals. Concentrations vary widely from ore to ore.	China	125,000[11]	Not available		China	36%	99 million in REO content
		Russia	2,470			CIS	19%	
		India	50			U.S.	13%	
		United States (processing of stockpiled ore at Mt. Pass, CA led to 2,150t REO[12])	0					
Lithium (in lithium carbonate equivalent/LCE)	Most lithium is recovered from subsurface liquid brines or from mining of lithium-carbonate rocks	Chile	38,720	Not available		Chile	76%	9.9 million in lithium content
		Australia	23,020			Argentina	8%	
		China	12,033			Australia	6%	
		United States	Withheld					
Cobalt	Primary cobalt (15%)	Ores, concentrates, or semi-refined materials:		Refined metals & chemicals:		DRC	51%	6.6 million in cobalt content
		DRC	25,000	China[13]	23,000	Australia	23%	
	Byproduct of nickel mining (50%)	Australia	6,300	Finland	8,900	Cuba	8%	
		China	6,200	Canada	4,900			
	Byproduct of copper mining (35%)	Russia	6,200	U.S.	0			
		U.S.	0					
Indium	Byproduct of zinc processing	Global: Not available		Metals, alloys, etc:		China	73%[14]	Not available
				China	300	Others	16%	
				South Korea	85	U.S.	3	
				Japan	60			
				U.S.	0			

Table 3-1. (Continued)

Tellurium	Byproduct of copper mining	Global: Not available U.S.: Withheld		Metals, compounds, etc:		U.S.	14%	22,000 in tellurium content
				Japan	40	Peru	10%	
				Peru	30	Others	73%[15]	
				Canada	20			
				U.S.: one refinery in Texas	data with held			
Gallium (Metal)	Most gallium produced as byproduct of treating bauxite; the remainder is produced from zinc-processing residues	China	59[16]	China	52[17]			See footnote 18
		Germany	35	Japan	85			
		Kazakhstan	25	U.S.: One company in Utah and one in Oklahoma	30			
		U.S.	0					

Sources: Eggert 2010, USGS 1994–2010a–e, Cobalt Development Institute 2009.

There are three primary criteria, among others, that determine the economic feasibility of a potential rare earth mine: tonnage, grade and the cost of refining the rare earth mineral. A mine may be economically viable (and therefore attractive to investors) if a low-grade (<5%) ore occurred with large tonnage and familiar mineralogy or if high-grade ore occurred with familiar mineralogy and moderate reserve tonnage. Globally, the four principal high-yield REE-bearing minerals are bastnäsite, monazite, xenotime and ion adsorption clays. A mineral deposit that does not fall in any of these four categories typically requires more metallurgical testing to establish the mineralogy and processing steps. The rare earth content of each deposit is essential to estimating the deposit's profitability. It determines how the ore will be processed and how complicated it will be to separate the rare earth elements from each other.[20] Of note is that nearly all rare earth deposits contain the radioactive material thorium and the cost of treating and storing thorium is an important factor in evaluating the economics of a mine. In general, each rare earth ore body is unique and requires a site-specific processing system. As a result, production costs vary from deposit to deposit based on ore content and mineralogy.

Much rare earth data is proprietary, yet public sources, including the United States Geological Survey (USGS), routinely publish data on the rare earth deposits.[25,26] Table 3-2 combines data from the USGS and others to summarize rare earth resources by source mineral type around the world. The table is not an exhaustive account of all known rare earth deposits.

The United States started producing rare earths out of Mountain Pass, California, in the mid-1960s and dominated global production of rare earths until 1984. However, China has been the world's leading rare earth producer since 1996 due to low production costs and valuable coproduction of iron ore at its principal rare earth mine in Inner Mongolia. The CIS, India and Brazil produce small amounts of REEs.[27]

Figure 3-1 shows production of rare earth oxides (REO) equivalent in the United States since 1990, compared to global production. The United States has substantial reserves of REEs, including small known amounts of heavy rare earth elements (HREEs). REE mining stopped in 2002, and as of December 2010 is limited to the reprocessing of rare earth ores that were stockpiled at the Mountain Pass mine.

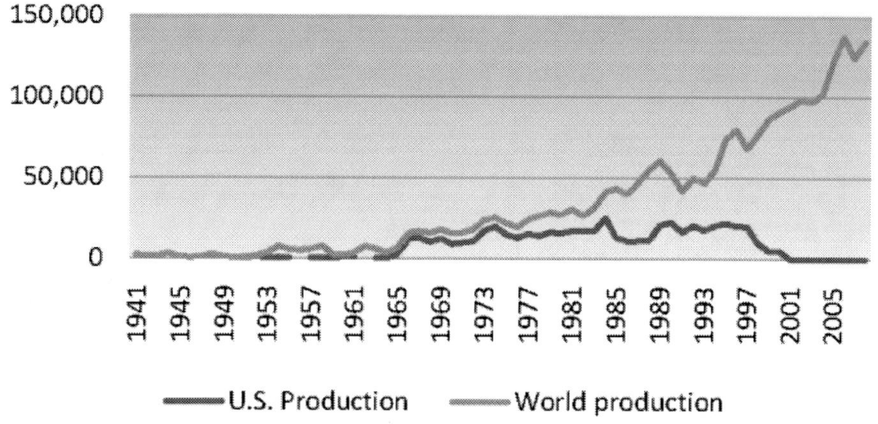

Source: USGS 2010g.

Figure 3-1. Historical production of rare earth oxides in the U.S. and the world in tonnes.

Table 3-2. Rare Earths Types and Contents of Major Contributing Source Minerals Supplying REEs to the Global Market (Percentage of Total Rare Earth Oxides)[21]

		LIGHT				MEDIUM			HEAVY							
TYPE	LOCATION (S)	Lanthanum (La)	Cerium (Ce)	Praseodymium (Pr)	Neodymium (Nd)	Samarium (Sm)	Europium (Eu)	Gadolinium (Gd)	Terbium (Tb)	Dysprosium (Dy)	Holmium (Ho)	Erbium (Er)	Thulium (Tm)	Ytterbium (Yb)	Lutetium(Lu)	Yttrium(Y)
Currently active:																
Bastnäsite	Bayan Obo, Inner Mongolia	23.0	50.0	6.2	18.5	0.8	0.2	0.7	0.1	0.1	0.0	0.0	0.0	0.0	0.0	0.0
Xenotime	Lahat, Perak, Malaysia	1.2	3.1	0.5	1.6	1.1	0.0	3.5	0.9	8.3	2.0	6.4	1.1	6.8	1.0	61.0
Rare earth laterite	Xunwu, Jiangxi Province, China	43.4	2.4	9.0	31.7	3.9	0.5	3.0	0.0	0.0	0.0	0.0	0.0	0.3	0.1	8.0
Ion adsorption clays	Longnan, Jiangxi Province, China	1.8	0.4	0.7	3.0	2.8	0.1	6.9	1.3	6.7	1.6	4.9	0.7	2.5	0.4	65.0
Loparite	Lovozerskaya, Russia	28	57.5	3.8	8.8	0.0	0.1	0.0	0.1	0.1	0.0	0.0	0.0	0.0	0.0	0.0
Various	India	23	46	5	20	4	0.0	0.0	0.0	0.0	0.0	0.0	0.0	0.0	0.0	0.0
Various	Brazil	N.A.														

Table 3-2. (Continued)

Possible to come online in the next 5 years:																	
Bastnäsite[22]	Mountain Pass, California, United States	33.2	49.1	4.3	12.0	0.8	0.1	0.2	0.0	0.0	0.0	0.0	0.0	0.0	0.0	0.1
Monazite	Mount Weld, Australia	26.0	51.0	4.0	15.0	1.8	0.4	1.0	0.1	0.2	0.1	0.2	0.0	0.1	0.0	0.0
	Eastern coast[23], Brazil	24.0	47.0	4.5	18.5	3.0	0.1	1.0	0.1	0.4	0.0	0.1	0.0	0.0	0.0	1.4
Apatite	Nolans bore, Australia	20.0	48.2	5.9	21.5	2.4	0.4	1.0	0.1	0.3	0.0	0.0	0.0	0.0	0.0	0.0
Fergusonite[24]	Nechalaco, Canada	16.9	41.4	4.8	18.7	3.5	0.4	2.9	1.8	0.7	0.0	0.0	0.0	0.0	0.0	7.4
Bastnäsite & Parisite	Dong Pao, Vietnam	32.4	50.4	4.0	10.7	0.9	0.0	0.0	0.0	0.0	0.0	0.0	0.0	0.0	0.0	0.007
Alanite & apatite	Hoidas Lake, Canada	19.8	45.6	5.8	21.9	2.9	0.6	1.3	0.1	0.4	0.0	0.0	0.0	0.0	0.0	1.3
Trachyte	Dubbo Zirconia, Australia	19.5	36.7	4.0	14.1	2.5	0.1	2.1	0.3	2.0	0.0	0.0	0.0	0.0	0.0	15.8
Not likely to be producing in the next 5 years:																	
REE thorium minerals	U.S. Rare Earths Lemhi Pass quadrangle, Idaho and Montana	7.0	19.0	3.0	18.0	11.0	4.0	11.0	0.5	4.0	0.5	0.2	0.2	0.5	0.2	20.9
Monazite	Nangang, Guangdong, China	23.0	42.7	4.1	17.0	3.0	0.1	2.0	0.7	0.8	0.1	0.3	0.0	2.4	0.1	2.4
	Eastern coast, Brazil	24.0	47.0	4.5	18.5	3.0	0.1	1.0	0.1	0.4	0.0	0.1	0.0	0.0	0.0	1.4
	North Capel, Western Australia	23.9	46.0	5.0	17.4	2.5	0.1	1.5	0.0	0.7	0.1	0.2	0.0	0.1	0.0	2.4
	North Stradbroke Island, Queensland, Australia	21.5	45.8	5.3	18.6	3.1	0.8	1.8	0.3	0.6	0.1	0.2	0.0	0.1	0.0	2.5
	Green Cove Springs, Florida, United States	17.5	43.7	5.0	17.5	3.1	0.8	1.8	0.3	0.6	0.1	0.2	0.0	0.1	0.0	2.5

Sources: USGS 2010j, Roskill 2010, IMCOA 2010 and GE 2010.

China currently produces at least 95 percent of global REEs (Roskill 2010). China introduced export quotas on REEs in 1999, citing the need for environmental management and resource conservation (see Table 3-3).[28] Between 2005 and 2009, REO exports fell by more than 20% from about 65,000 tonnes to about 50,000 tonnes.[29] In July 2010, China imposed the tightest quota thus far, leading to a 40% annual drop of exports.[30] This latest set of export quotas were non-element specific and applied to all exports of REEs, which in turn led to price spikes for the lower valued light rare earth elements (LREEs) as traders favored exports of the more valuable HREEs. China's Ministry of Industry and Information Technology may propose additional measures on some REEs in the 12th Five-year Plan for Rare Earth Industry sometime in 2010 (*Business China* 2010).

Additionally, it is estimated that another 20,000 tonnes are illegally exported from China bringing total production to approximately 145,000 tonnes of REO (China Daily 2009, Kingsnorth 2010).

Reserves and Production of Lithium

Currently, economically viable lithium resources are found mainly in South America.[32] Globally, it is more economic to extract lithium in continental brines than in hard rocks or spodumene deposits. Among the continental brines, South American brines hold the most favorable lithium chemistry and are currently most economic to mine. However, lithium is found in many countries around the world, including China (continental brine) and the United States (continental brine, oil field brine and geothermal).[33] Currently there is also excess production capacity of 46% and additional lithium mines could come on line if greater demand further increased prices.[34]

Table 3-3. China's REE Export Quotas and Demand from Rest of the World (ROW): 2005–2010

Export Quotas (tonnes REO)		Change from Previous Year	ROW Demand (tonnes)	ROW Supply[31] (tonnes)
2005	65,609	-	46,000	3,850
2006	61,821	-6%	50,000	3,850
2007	59,643	-4%	50,000	3,730
2008	56,939	-5%	50,000	3,730
2009	50,145	-12%	25,000	3,730
2010	30,258	-40%	48,000	5,700–7,700

Sources: Kingsnorth 2010, Koven 2010, Hatch 2010.

In 2009 Chile, Australia and China together accounted for 78% of global lithium production.[35] Globally, the biggest suppliers of lithium are Chemetall and SQM (Chilean), Tailson Minerals (Australian), FMC (American) and three mining companies in Sichuan, China (Roskill 2009; Baylis 2009). Chile accounts for 41%, Australia for 24% and China accounts for close to 13% of current global lithium production.[36] China consumes most of its domestically produced lithium and is developing capacity to produce high-purity lithium compounds.[37]

The United States produced lithium minerals from hard rock ores until 1997, when the spodumene mine in North Carolina closed due to its inability to compete with South American brines. The United States currently has only one active lithium brine operation in Nevada.[38] Two U.S. companies produce and export a large array of value-added lithium materials produced from domestic and South American lithium carbonates (USGS 2010a).

Reserves and Production of Cobalt

Currently, most cobalt is produced as a byproduct of nickel and copper mining. The Democratic Republic of the Congo (DRC) produces 40% of global cobalt as a byproduct from copper mining and artisanal mining. DRC holds about half of the world's identified cobalt reserves.[39] DRC experienced a recent civil war[40] and mining contracts awarded during the conflict under a transitional government were renegotiated. The country remains politically unstable and is one of the lowest ranked countries on the global Policy Potential Index.[41] Nevertheless, the Cobalt Development Institute projects that DRC's dominance over cobalt production will continue to grow in the near future (Cobalt Development Institute 2010).

Cobalt produced as a major product (ores, concentrates and intermediate materials) occurs mostly in Morocco, but also via artisanal mining and recovery from previously stockpiled intermediate materials in DRC. [42]

The leading global producers of refined cobalt are China (39%), Finland (15%) and Canada (8%). China refines cobalt based on the primary cobalt imported from DRC (USGS 2010b). DRC used to be a leading cobalt refiner and will likely increase refinery production again.[43]

TEXT BOX 3-2

Coproducts and Byproducts

Unlike industrial materials such as copper and zinc that are produced as major products, the materials addressed here are minor metals (including specialty, precious, and "rare" metals) chiefly as coproducts or produced byproducts.

All REEs appear naturally in different combinations within a single mineral form, making it infeasible to mine for individual REEs. The packet of individual REEs can instead be considered as coproducts.

Other minor metals such as indium, tellurium, gallium, and most cobalt are primarily produced as byproducts of other mining operations. The availability of the byproduct is greatly influenced by the market dynamics of the major product.

Major product	Co- or byproduct
Nickel, copper	Cobalt
Copper	Tellurium
Zinc	Indium, gallium
Higher profit rare earth elements (Nd)	Lower profit rare earthelements (La, Ce, Sm)

The United States has not mined cobalt since 1971 and has not refined cobalt since 1985. In recent decades, the United States has been recovering negligible amounts of cobalt from Missouri's lead ore and from the mining and smelting of platinum group metals (PGMs) in Montana. Imports, secondary sources (i.e., recycled scraps and spent materials) and stock releases have been the United States' major sources of cobalt. [44] Several projects are under development to expand cobalt production in the United States; the Idaho Cobalt Project plans to produce cobalt as a primary product in 1 to 2 years and two other projects will produce cobalt as byproducts—the Eagle Project nickel-copper mine in Michigan and the NorthMet Project copper-nickel-PGM mine in Minnesota (USGS 2010b).

Reserves and Production of Gallium

Gallium exists in very small concentrations in ores of other metals, mostly bauxite and zinc. Most gallium is produced as a byproduct of treating bauxite to extract aluminum and the remainder is produced from zinc-processing residues. World resources of gallium in bauxite are estimated to exceed 1 billion kilograms, and a considerable quantity could be present in world zinc reserves. However, only a small percentage of this metal in bauxite and zinc ores is economically recoverable globally. An estimate of current reserves of gallium comparable to the definition of reserves of other minerals thus cannot be made (USGS 2010c).

Assuming that the average content of gallium in bauxite is 50 parts per million (ppm), U.S. bauxite deposits, which are mainly sub-economic resources, contain approximately 15 million kilograms of gallium. Some domestic zinc ores also contain as much as 50 ppm gallium and, as such, could be a significant resource (USGS 2010c).[45]

Table 3-4. U.S. Demand Characteristics of Key Materials

	Principal end uses (in volume)[49]		Recycling and stock info	Reported use (*tonnes*) and U.S. consumption as percentage of global	U.S. import sources (top three only)		U.S. net import dependence (percentage of reported consumption)
Rare earth oxide (REO)	Metal alloy	29%	Small amount recovered from spent permanent magnets	15,500 (2008)	Metals, compound, etc.:		100%
	Electronics	18%					
	Chemical catalysts	14%		12.5% (2008)	China	91%	
	Phosphors	12%			France	3%	
	Catalytic converters	9%	Producer and processor stock info withheld		Japan	3%	
	Polishing & ceramics	6%					
	Magnets	5%					
	Oil refining catalysts	4%					
Lithium (LCE)	Ceramics & glass	31%	Recycling insignificant but increasing; Producer stock info up to 1999 but withheld	6,280 (2009)	Chile	63%	>50%
	Batteries	23%			Argentina	35%	
	Greases	10%		7% (2009)	China	1%	
	Air treatment	5%					
	Continuous casting	4%					
	Other	27%					
Cobalt (metal)	Batteries	25%	Cobalt from purchased scrap met 24% 2009 reported consumption; Industry year-end stock info available	7,000 (2009)	Norway	19%	75%
	Super and other alloys	25%		12 % (2009)	Russia	17%	
	Catalysts	10%			China	12%	
	Magnetic alloys	7%					
	Carbides	13%					
	Other chemical and ceramic uses	20%					
Indium (metal)	Coatings (ITO for LCDs)	80%	Recovery from manufacturing wastes mostly in China, Japan and South Korea; recycling could rise significantly in U. S.; recovery from tailings viable	120 (2009)	China	40%	100%
					Japan	19%	
				N.A.	Canada	18%	
	Others (alloy and solders; semiconducting compounds for LEDs; solar materials)	20%					
Tellurium (metal)	Metallurgy	45%	Little or no scrap to extract secondary tellurium	50 tonnes[50]	China	43%	Not reported
	Solar cells	25%		~10% (est. based on Eggert (2010)	Belgium	24%	
	Rubber processing & synthetic fiber	20%	Stock info withheld		Canada	18%	
	Electronics	10%					
Gallium (metal)	Integrated circuits	67%	World gallium recycling capacity at 42% of 2009 production capacity[51]	29 (2008)	Germany	24%	99%
	Optoelectronic devices (cell phones, backlights, flashes)	31%		25%[52]	Canada	20%	
	Other	2%			China	16%	

Sources: USGS 2009-2010a-e, CDI 2009, Eggert 2010.

Reserves and Production of Indium

Global primary production of indium is widely distributed because indium is a byproduct of a number of industrial minerals. Currently economic reserves of indium are concentrated in China (73%), Peru (4%), the United States (3%) and other countries (16%). Indium is recovered almost exclusively as a byproduct of zinc production. Significant quantities of indium are also contained in copper, lead and tin ores, but most deposits are sub-economic. Globally, half of indium refining takes place in China, followed by South Korea (14%) and Japan (10%). China implements export quotas on indium and indium products. The 2009 Chinese indium export quota was 233 tonnes, a 3% decline from the 2008 indium export quota of 240 tonnes. China is anticipated to continue to tighten its indium export quota to meet a growing domestic demand. Indium can be reclaimed from spent indium-tin-oxide (ITO) sputtering targets and cuttings generated during ITO target processing. Technology has been developed to recover indium directly from liquid crystal display (LCD) glass. Indium can also be recovered from tailings when the price is high.[46]

Reserves and Production of Tellurium

Australia, Belgium, China, Germany, Kazakhstan, the Philippines, Russia and the UK hold around 73 percent of global tellurium reserves. Most tellurium is recovered from processing copper deposits. With increased global concern over tellurium supply, companies are investigating other potential resources, such as gold telluride and lead-zinc ores with higher concentrations of tellurium. These ores are not currently included in the estimates of world tellurium resources (USGS 2010e).

3.2. Demand

The two major drivers of demand for mineral commodities are the rate of overall economic growth, (stable or decline) and the state of development for principal material applications (e.g., clean energy technologies). Demand for key materials in clean energy technologies compete for available supply with demand for the same materials in other applications.[47]

Several additional pieces of information are helpful for understanding demand-supply mismatches for the United States: domestic demand as a share of global demand, import dependence[48], stock releases, substitutes, recycling and greater material use efficiency. Import dependence and supply risks in general should be examined over the entire supply chain. A case in point is that around 40% of global cobalt mine production occurs in DRC whereas only around 2% of global refining of cobalt into metal occurs in DRC (USGS 2010f). This indicates that countries importing refined cobalt can still be indirectly dependent on cobalt from DRC, a politically unstable country.

Historical Demand for Rare Earth Elements[53]

Recently, REE consumption has seen large regional growth mainly due to the growth of advanced technology and clean energy technology sectors. In China and globally, REEs have experienced fast growth in advanced technology sectors including luminescent (phosphors), magnetic, catalytic and hydrogen storage technologies.[54] The demand by clean energy technology sectors is largely a result of the ramp-up of clean energy technology manufacturing and use by the United States, other Organization for Economic Co-operation and Development (OECD) nations and China. Magnets dominated REE usage by weight in 2008, with catalysts claiming the second-highest usage and metal alloys accounting for the third highest (Kingsnorth and Chegwidden 2010). REE consumption has grown most rapidly in China China's 2005 REO demand exceeded half of global demand for the first time and more than tripled in absolute terms between 2000 and 2008 (Chen 2010).

The United States was responsible for around 12% of global rare earths demand (combined demand of REOs and REO equivalent of chlorides, compounds, alloys and metals) in 2009. U.S. demand for rare earths has drastically changed over the last 30 years. To meet domestic demand, the United States increasingly relies on imports of rare earth metals, alloy, compounds, oxides, among other forms of rare earth containing materials. At the same time, the United States has been exporting rare earth metals, alloys and compounds and in 2009, became for the first time a net exporter of REO equivalents.[55] Figure 3-2 shows U.S. demand for all REO equivalents since 1970, along with its historical demand for three other key materials.

Historical Demand for Lithium

Global lithium consumption is driven mainly by the growth of rechargeable lithium batteries and the strong economic growth of the emerging economies. The production of rechargeable lithium batteries grew by 25% per year between 2000 and 2007. Lithium used for battery production now accounts for more than 20% of total lithium consumption, compared to 6% in 2000 (Roskill 2009). Despite the 2009 economic downturn which led to lower demand, worldwide exploration for lithium proceeded, driven by strong economic growth in emerging markets. The major industrial uses for lithium—ceramics, glass, batteries and lubricating greases—have also benefited from robust economic growth in emerging markets.[56]

Figure 3-2 shows U.S. historical demand for lithium since 1970. From 2003–2007, lithium demand increased by about 8% but growth slowed to just 4% in 2008 due to the economic downturn. The United States has been mostly dependent upon lithium imports since the late 1990s, with its current imports coming chiefly from Chile and Argentina.[57] The United States also consumes recycled lithium, though not at a significant level.[58]

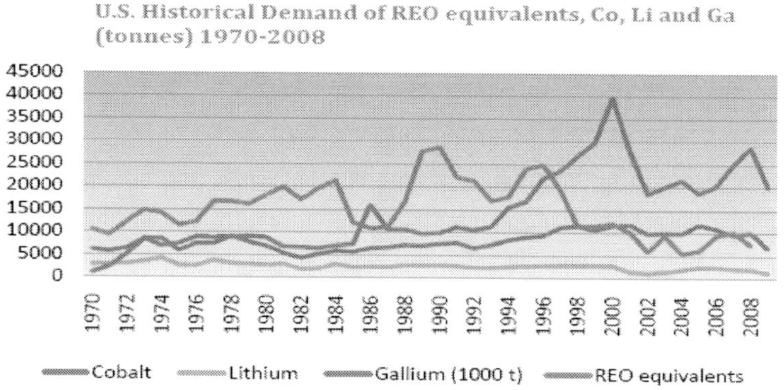

Source: USGS 2010g.

Figure 3-2. U.S. historical demand for REO equivalents, Li, Co and Ga.

Historical Demand for Cobalt

Cobalt demand is driven by general economic conditions and traditional demands from industries such as the superalloy sector and rechargeable battery manufacturing for small consumer device applications. The superalloy sector includes manufacturers of turbine engine parts for jet aircraft and land-based energy-generating turbines (USGS 2010b). Similar to other materials discussed in this chapter, cobalt consumption has seen a rapid rise in China (Cobalt Development Institute 2010).

The USGS estimates that in 2009, 49% of the cobalt consumed in the United States was used in superalloys, mainly in aircraft gas turbine engines; 9% in cemented carbides for wear-resistant applications and cutting; 15% in various other metallic applications; and 27% in a variety of chemical applications. Figure 3-2 on the previous page presents U.S. demand for cobalt since 1970.

The United States is currently about 75% import dependent upon cobalt coming from Norway, Russia and China. Recycled cobalt from purchased scrap helped the United States meet 24% of its domestic consumption in 2009 (USGS 2010b).

Historical Demand for Gallium

Gallium demand is growing in several applications including light-emitting diodes (LEDs) used for liquid crystal displays in televisions and notebook computers and solar cells. In addition, its material intensity in solar cells has been declining thanks to efficiency improvements. Electronic components have represented about 98% of U.S. gallium consumption since 2003. In 2009, about 67% of the gallium consumed was used in integrated circuits (ICs). Optoelectronic devices, which include laser diodes, LEDs, photodetectors and solar cells, represented 31% of gallium demand.[59] The remaining 2% was used in research and development, specialty alloys and other applications. The global economic downturn hurt LED markets, although emerging LED market segments, such as for LCDs in televisions and notebook computers, still showed growth. At the same time, record-making solar cell efficiencies are reducing the need for gallium, among other materials, in making thin film solar cells (USGS 2010c).

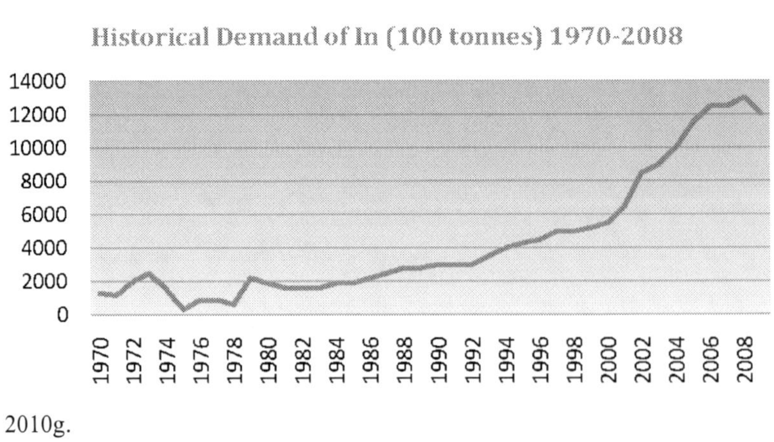

Source: USGS 2010g.

Figure 3-3. Historical demand for indium.

The United States represents about 25% of the global annual consumption of gallium. Since 1982, the United States has been dependent chiefly on imports for meeting its annual gallium demand. 60 The United States currently imports gallium from Germany (24%), Canada (20%), China (16%) and the Ukraine (12%).[61] Figure 3-2 shows U.S. historical demand trends for gallium since 1970.

Substantial quantities of new scrap are being reprocessed, although data on the amount and usage are not available.

Gallium is used in a promising new category of photovoltaic (PV) solar cells based on an alloy of copper, indium, gallium and selenium. To date, this type of solar cell accounts for a very small share of the solar market (USGS 2010c).

Historical Demand for Indium

Global demand for indium exploded about a decade ago due to its use in flat-panel displays, television sets, computer monitors and smart phones. Thin films of indium-tin-oxide (ITO) form an integral part of all three aforementioned items. Manufacturers of ITO thin films responded to high indium prices by recycling indium previously discarded as manufacturing waste. However, the amount of indium in each flat-panel product has largely remained the same.

Indium also enters the promising new category of PV solar cells referred to at the beginning of this page, along with copper and selenium. Again, this type of solar cell still accounts for a very small share of the solar market (USGS 2010d).

The United States has been 100% dependent on imports for indium since 1972, with current indium imports coming from China, Japan and Canada.[62] In terms of secondary sources, indium is recycled from manufacturing wastes in China, Japan and the Republic of Korea—the countries in which most ITO production occurs. According to USGS, recovering indium from the tailings of zinc mining is possible when the price is high (USGS 2010d). Figure 3-3 presents the U.S. historical demand for indium since 1970.

Historical Demand for Tellurium

Tellurium is used in the production of high-performance solar cells. Cadmium telluride is one of the most promising thin-film photovoltaic compounds for power generation, achieving some of the highest power conversion ratios. Despite a drop in the overall demand for solar cells in 2009 due to the economic downturn, the demand for cadmium telluride solar cells continued to rise.

Information about U.S. tellurium demand and imports is difficult to obtain without disclosing proprietary data. However, it is known that the United States' principal tellurium import sources are China, Belgium and Canada. Several materials, including selenium, can replace tellurium in most of its uses, but usually with losses in production efficiency or product characteristics. There is little or no scrap from which to extract secondary tellurium because the uses of tellurium are nearly all dissipative in nature. A small amount of tellurium is recoverable from scrapped selenium-tellurium photoreceptors employed in older photocopiers in Europe (USGS 2010e), but this has decreased over time.

3.3. Prices

Supply risks, at least in the short-to-medium term, are less associated with the prospect of increasing prices because in most cases the cost of these elements is a small part of the final product manufacturing cost. However, in the last 6–12 months the price of many rare earth elements has increased by approximately 300–700%, which in some cases has had a more significant impact on the price of the final product (Lynas Corp. 2010).[63]

Main Factors driving the prices of key materials

The 1980s and 1990s were a time of over-supply of minerals due to decreased demand for commodities following the two oil shocks of the 1970s. Supplies of many minerals rose in the 1980s when major mines opened, particularly in Latin America and Southeast Asia and in the 1990s when large volumes of metals entered global markets after the collapse of the Soviet Union. The lengthy period of over-supply hindered the price of any mineral from rising above its short-term marginal cost for a sustained period (Humphreys, forthcoming).

In contrast to the 1980s and 1990s, price concerns are more salient today as the market has moved from one where there was oversupply of minerals to one where there is more concern about undersupply. The decline in the value of the U.S. dollar has contributed to higher metal prices when they are presented in terms of U.S. dollars. As most mineral markets are priced in dollars, the declining value of the dollar leads to higher prices for many metals. A sustained demand boom for many commodities, mainly driven by Asia's rapid industrialization, has followed the recovery from the 2001 economic recession. The rapid rise in demand, particularly from China and the decline in the value of the U.S. dollar have in many cases driven the prices to a historical high though not necessarily led to a sustained high price level.

Despite the rising demand and the historical high prices reached by many commodities, mine capacity expansions and new mine production capacity have not kept pace. Some factors behind delayed development are region specific, though a major cause has been the generally rising costs of metal production and production capacity expansion.[64] The economic crisis in 2009, which made it more difficult for projects to get financing, caused further delays.

Overall, the price of minerals is driven by multiple physical, financial and political factors. When deciphering price data and trends, it is helpful to know whether there is a market surplus or deficit and the extent of the imbalance. Physical parameters (e.g., stock changes, closures of old mines and the start-up of new ones) are in turn influenced by general economic conditions and financial forecasts (e.g., inflation and exchange rates, that inform investor sentiments. Unanticipated shocks— such as monopolistic or oligopolistic pricing (e.g., export quotas), geopolitics and natural disasters— also play a role in affecting physical and financial parameters.

To understand the price behavior and volatility of key materials, it is also important to examine the ways in which these commodities are bought and sold, in conjunction with whether they are produced as a co- or byproduct of other specialty metals (e.g., REEs) or a by-product of a major metal. Both aspects influence the price behavior and volatility of a mineral. The influences of these factors can be gleaned from a comparison between the historical price trends of commodities produced as a byproduct of metals traded on major exchanges and commodities mainly transacted through bi-lateral contracts.

Negotiated pricing and metal exchanges

Most rare, precious, minor and specialty metals and their alloys are traded through bilateral contracts based on negotiated pricing between parties. The fragmented nature of some of these markets and the remoteness of some producers has resulted in traders playing a dominant role. Regionally, traders account for a large part of the specialty metal supply coming out of regions such as China, the former Soviet Union and Africa. The nature of the process limits price disclosure in these markets and the prices of specialty metals quoted by traders and consultants vary widely in their reliability (Humphreys, forthcoming). Though not considered a minor mineral, lithium prices have been available mainly through trade journals.

Several of the key materials considered in this Strategy are produced as by-products of nickel, copper or zinc refining. These three major metals are some of the most economically important, non-ferrous metals. They are typically traded and priced on metal exchanges such as the London Metal Exchange (LME) and the New York Mercantile Exchange (Nymex). As of 2008, it was possible to trade all three of these primary metals on the LME. The Nymex offers contracts in copper. Trading via metal exchanges indicates a larger volume of transactions compared to negotiated trading. The large scale of trades through the LME or the Nymex is also due in part to the opportunity for hedging and speculating. As a result, price data on metals transacted through metal exchanges are large in volume and available in most areas of the world.

Table 3-5. Purchase Option and Source of Price Information for Key Materials of Concern

Minerals/Metals	Purchase option	Source of price info
Rare earth elements	Negotiated purchase, not traded on metal exchanges and therefore no spot or future market; however, illegally-traded REEs are sold through less formal channels and may possibly be sold on the spot markets	Trade journals, based on information from producers, consumers and traders
Cobalt (most), gallium[65], tellurium, indium, lithium	Negotiated purchase, not traded on metal exchanges and therefore no spot or future market (except for indium and small amount of cobalt)	Trade journals, based on information from producers, consumers and traders
Cobalt (small share)	Cobalt became tradable on LME in February, 2010. Producers registered with LME for trading certain brands of cobalt so far maintain a combined warehoused amount of 115 tonnes, which is small compared with the 60,000 tonnes global cobalt market; spot market	Information available globally from the exchange
Nickel (Ni), copper (Cu), zinc (Zn)	LME, copper is also traded on COMEX (part of NYMEX)	Information available globally from the exchange; trade journals

Sources: Humphreys, forthcoming; USGS.[66]

Table 3-5 summarizes the purchase options and price information sources for the key minerals considered in this Strategy. Among these materials, cobalt, indium and tellurium are byproducts of metals (i.e., nickel, copper and zinc) that trade on metal exchanges. By contrast, rare earth oxides and rare earth metals are typically traded through long- or short-term bilateral contracts.

Joint production and prices

Byproduct or coproduct material availability is influenced by the commercial attractiveness of the associated major product (see Text Box 3-2). For example, if the price of the major product falls, less mining of ore containing the major product will occur and, as a result, there will be less byproduct available to recover. Or, if the price of the byproduct rises, such a price rise alone may have little or no impact on the amount of major product ore that is mined and thus the amount of the byproduct may remain unchanged, despite the higher price.

Commodities that do not trade on metal exchanges or do not have a market on which to be sold to a buyer of last resort (e.g., rare earths and lithium) face greater pressure to respond to market conditions by cutting output when a producer cannot find a buyer or storage space. In such markets, volume change is the common mechanism used rather than price adjustments or price swings (Humphreys, forthcoming).

Historical price data

The following are historical price trend data and accompanying information for the materials examined in this report (except for tellurium due to unavailable data). For each price trend, a description of the factors and reasons behind major price shifts is given. Generally, each minor metal exhibits somewhat different price trends due to peculiarities in each market (e.g., indium dominated by ITO demand in electronics and flat-panel displays, cobalt prices in 2008-2009 reflecting supply disruptions in Africa and Canada, etc.). This is in sharp contrast to price behavior of major metals (such as Ni and Cu) which is more heavily influenced by overall gross domestic product growth and macroeconomic conditions. The fortunes of the minor metals are more closely tied to a small number of end-use sectors and often a small number of producers. As a result, minor metal markets are more fragile than the major metal markets (Eggert 2010).

Figure 3-4 illustrates the historical average prices of individual rare earth oxides between 2001 and 2010. This period covers the 2001 recession, which had lingering effects until 2003, and the 2008– 2009 recession. Two things to note are that the heavier rare earths (e.g., dysprosium, terbium and europium) are relatively more expensive, and that prices have risen fairly steadily since 2003 due to China's rising domestic demand and escalating export controls. The price jumps from 2009–2010 can perhaps be attributed to a reduction in China's rare earth export quota. The export quota which is for total rare earth exports, resulted in higher prices for REO exports. This led to an unexpected fall in China's export of LREEs which are generally lower priced. As a result of the greater scarcity of light rare earths, the price of LREEs rose much more than the HREEs. Rare earth oxide and rare earth metal prices track closely, with the prices for metals always higher (though relatively more so for some rare earth elements than others) (British Geological Survey 2010).

Figure 3-5 tracks the price of cobalt during a similar period. A market surplus of cobalt started in 1996 and lasted until the early 2000s, when a strong demand in 2004 led to a spike in cobalt prices.[67] Around that time, health, safety and environmental issues started to become increasingly significant to the market for metals such as cobalt.[68] Cobalt prices trended downward from 2005– 2007, reflecting an adequate supply of refined cobalt overall. Strong demand in 2007 was followed by projections of several new mine or refinery projects coming online in 2008; however, the world's available refined cobalt fell in mid-2008 as a result of the industry's response to low prices and reduced demand. The responses included the closure of a Zambian refinery, cutbacks at numerous nickel operations and some copper-cobalt operations in DRC and the delayed startup of proposed brownfield and greenfield projects. More recently, the global economic downturn caused cobalt prices to decrease. As economic conditions improved, cobalt production has increased to levels where there is enough supply to meet increasing demand. [69]

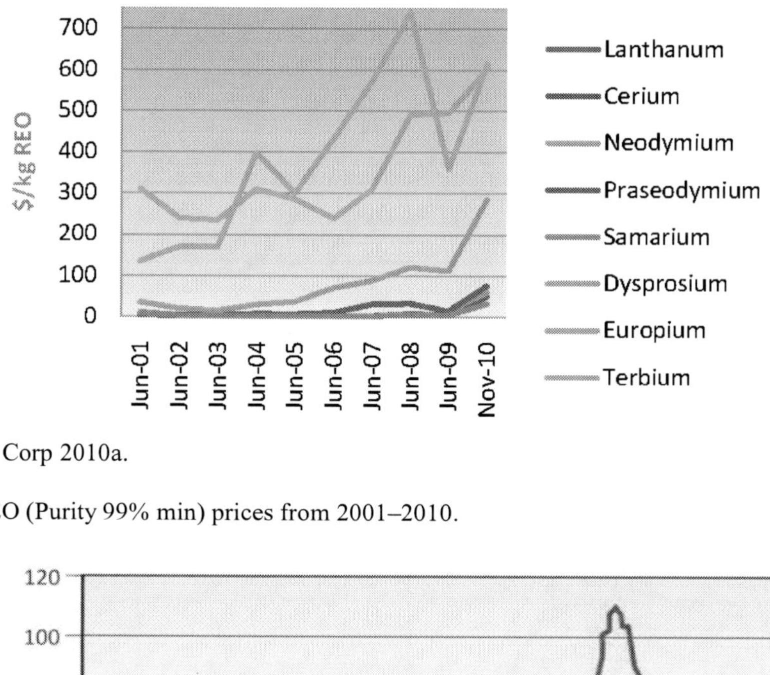

Source: Lynas Corp 2010a.

Figure 3-4. REO (Purity 99% min) prices from 2001–2010.

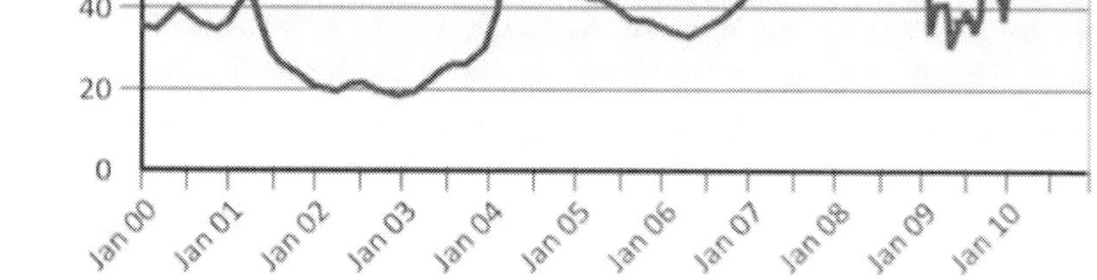

Source: Arnold Magnetic Technologies 2010.

Figure 3-5. Cobalt prices from January 2000 to January 2010.

Figure 3-6 provides detailed historical price information for lithium carbonates, illustrating price trends driven by the opening of new brine mines and closing of old spodumene mines as well as continuous regional demand growth. In the early 1990's, the United States was the largest producer and consumer of lithium minerals and compounds worldwide. In the early 1990s the U.S. Department of Energy also sold about 37,200 tonnes of excess lithium material from the thermonuclear weapons programs of the 1950s and 1960s.[70] In 1997, the U.S. closed down its last spodumene mine in North Carolina and lithium carbonate production from hard rock ores in the U.S. ended. In contrast, a second lithium brine operation in Chile completed its full year of operation in 1997, with a higher production of lithium carbonate than was initially expected. The increased production from SQM in Chile significantly lowered the lithium price and eliminated their spodumene competition, allowing the company to gain substantial market share. The recent price movements beginning in 2005-2006 are mainly due to the reality that there are only a few lithium producers in the world. The mid-2000s saw bad weather intervening with Argentina's lithium production. Following that was a period of insufficient production to meet rapidly growing demand. Prices started to level off and then decrease slightly by early 2008 when surplus Chinese lithium began to hit the market and balance demand and supply. Lithium prices remained stable even when the economic downturn hit in late 2008 and throughout 2009. Lithium prices only started to decrease beginning with new contracts in early 2010, due to SQM lowering its prices by 20%, and other producers following suit to some degree.[71]

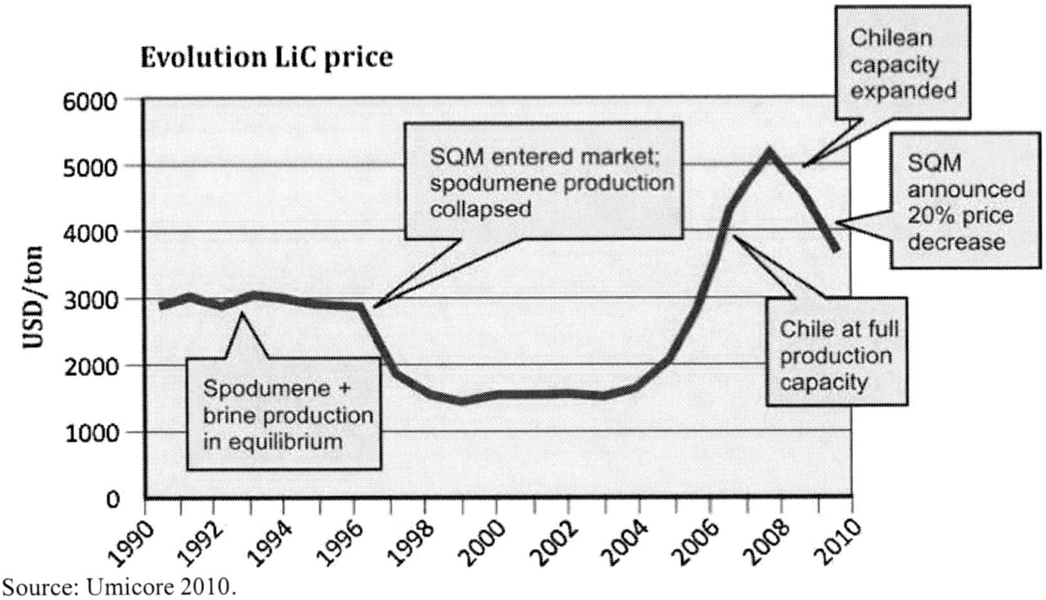

Source: Umicore 2010.

Figure 3-6. Lithium prices from 1990–2010.

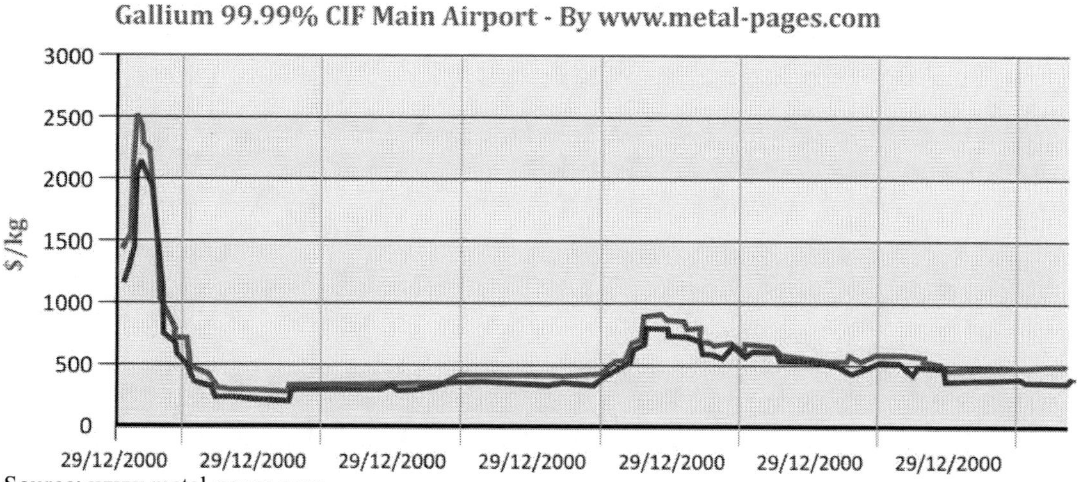

Source: www.metal-pages.com.

Figure 3-7. Gallium prices in the European market from 2000–2010.

Figure 3-7 shows historical gallium price trends in the European market, which follows world gallium prices closely. Gallium supplies were tight in 2000 because of continuously increasing demands for wireless communication products. Until early 2001, supply remained tight and the price for high-purity gallium reached $2,500 per kilogram (kg). By mid-2001, gallium spot prices dropped to about $1,000 per kg, still higher than the average selling price of $500–$600 per kg. The U.S. economic slowdown resulted in a decline in the cellular telephone market, which had been principally responsible for the growth in gallium consumption in the previous few years. In 2002, one of the two gallium refiners in the United States exited the business due to the slump in demand by the telecommunications industry. Prices for low-grade (99.99%-pure) gallium increased in the first half of 2007 and producers in China claimed that a shortage of supply was the principal reason for the increase in prices.[72]

Figure 3-8 shows historical indium prices. In the early 2000s, expanding LCD manufacturing in Asia was more than matched by an adequate supply and highly efficient processing. Despite a strong increase in LCD production, the ready availability of low-priced indium from China forced world prices down. In 2003 and 2004, reduced production from mines that generated byproduct indium and the closure of several smelters—due to environmental problems—created the perception that supplies of indium from China would decrease and drove world indium prices to historic highs. The indium price continued its remarkable rise into the fall of 2005, driven by continued strong sales of flat-panel displays and other LCD products that increased global consumption of ITO. Global secondary indium production increased significantly during 2005–2007 and accounted for a greater share of indium production than primary production by 2007. Global ITO demand continued to rise, leading to some price spikes caused by supply deficits and the indium supply's heavy dependence on the strength of the zinc market (USGS 1994–2010d).

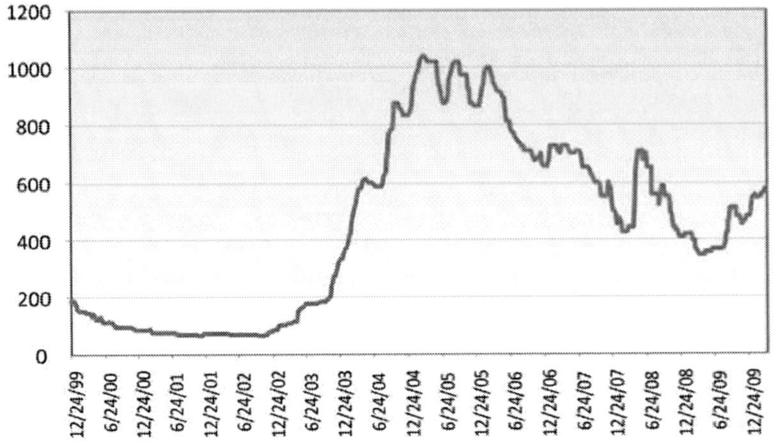

Source: metal-pages.com.

Figure 3-8. Indium prices from 2000–2010 (99.99% pure metal in $/kg).

3.4. Conclusions

This chapter addressed the historical supply, demand and price data for materials important to clean energy technologies. These materials have already seen a rise in demand driven by one or more uses, and the United States is heavily import-dependent for most of them. These materials are also predominantly co-produced with other metals, resulting in additional supply risks. Although there are secondary sources for some of these materials from recovered scrap and stock releases, such sources still meet a relatively small share of U.S. and/or global demand (except for indium). Among the materials of interest, other than the rare earths, the United States appears to have some level of diversity in terms of import sources. However, the complex supply chain could still lead to an indirect reliance on these less stable sources. Information about future supply and demand, as well as an assessment of the potential mismatch between supply and demand for each material, is presented in Chapter 7.

4. Current DOE Programs

Several U.S. Department of Energy (DOE) data and information programs, research and development (R&D) programs and financial instruments address rare earths and other key materials. Current programs focus on the component and end-use technology stages of the supply chain and address both the economic and the innovation dimensions of the clean energy sector.

4.1. Data and Information

Data and information can inform economic policies and R&D priorities both inside and outside DOE. The Energy Information Administration (EIA), an independent agency within the DOE, collects, analyzes and disseminates independent and impartial energy information. EIA is the nation's premier source of energy information. The EIA has been studying the supply of rare earth materials, the consumption of those materials in clean energy technologies and impacts of rare earth use on technology cost and performance. EIA plans to develop improvements to existing surveys to monitor the deployment of technologies that use the materials.

4.2. Research and Development

DOE R&D programs supporting scientific and technological innovation range from basic research to large-scale technology deployment. DOE supports programs from low-risk, evolutionary projects to high-risk, high-payoff experiments. These programs span the entire energy innovation pipeline but are closely connected (Figure 4-1). DOE also supports R&D addressing specific materials and alternatives across the supply chain. In Fiscal Year (FY) 2010, the Office of Science, the Office of Energy Efficiency and Renewable Energy (EERE) and the Advanced Research Projects Agency-Energy (ARPA-E) together provided approximately $15 million for research on rare earth materials and possible substitutes for magnets. An additional $35 million was spent by ARPA-E on next generation battery technologies that don't require rare earths.

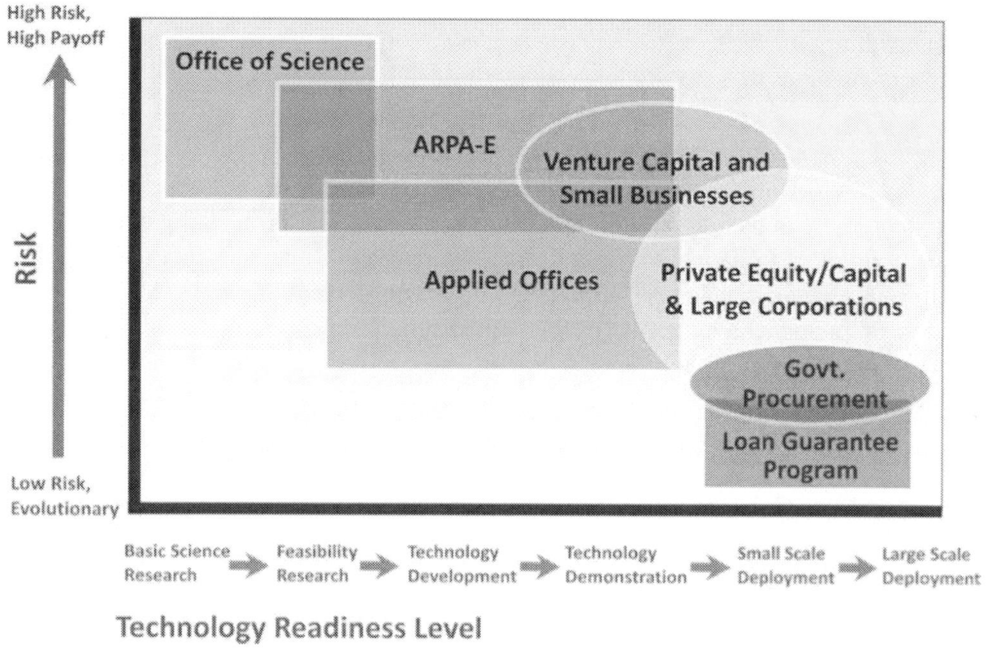

Figure 4-1. Energy innovation pipeline.

Office of Science

At the basic science end of the pipeline, the Materials Sciences and Engineering (MSE) Division of the Office of Basic Energy Sciences supports broad-based, fundamental materials research. MSE seeks to illuminate the atomic basis of materials properties and behavior and improve materials performance at acceptable costs through innovative design, synthesis and processing. This research was funded at a level of about $5 million/year in FY2010..

Most of the supported work has been performed at Ames Laboratory. This work includes materials synthesis and processing, phenomenological behavior investigations and characterization. The main emphasis is on rare earth materials that change temperature, shape or electrical resistance upon exposure to a magnetic field. The research focuses on the synthesis of highest quality polycrystals and single crystals, advanced characterization methods, especially neutron and magnetic X-ray scattering and first principles modeling. The ultimate goal of the research is to understand and control the responsiveness of materials that are sufficiently complex to facilitate control at length scales ranging from electronic interaction distances to atomic and microstructural scales.

A key component of the Ames Laboratory program is the Materials Preparation Center (MPC). The MPC was established in 1981 to provide high purity metals (including the rare earths, uranium, thorium, vanadium, chromium); and intermetallics, refractory, inorganic compounds and specialty alloys; none of which are available commercially in the required purity or form/shape needed by the requestor on a cost recovery basis. The Center is focused on establishing and maintaining materials synthesis and processing capabilities crucial for the discovery and development of a wide variety of use-inspired, energy-relevant materials in both single crystalline and polycrystalline forms, spanning a range of sizes with well-controlled microstructures.

The Office of Science also supports the development and validation of models to theoretically identify promising structures and compositions in order to synthesize rare earth materials with optimum properties. This Materials by Design approach also is informing the search for compounds that are suitable rare earth substitutes.

Advanced Research Projects Agency-Energy

Moving along the pipeline to applied research via feasibility research, technology development and demonstration, ARPA-E supports two initial projects totaling $6.6 million specifically targeted to developing substitutes for rare earth magnets. The goal of this $4.4 million project is to develop materials to allow the United States to fabricate the next generation of permanent magnets (PMs) with magnetic energy density (maximum energy product) up to two times higher than the current value of the strongest commercially available neodymium-iron-boron (Nd-Fe-B) magnets. If successful, this project will lead to cheaper, more energy-efficient, more power-dense magnets for deployment in a wide range of clean energy technologies.

In another ARPA-E project, General Electric Global Research (GE) is developing next-generation permanent magnets with a lower content of critical rare earth materials. For the $2.2 million project, GE is developing bulk nanostructured magnetic materials with a dramatic increase in performance relative to state-of-the-art magnets. These new magnets will increase the efficiency and power density of electric machines while decreasing dependence on rare earth minerals. If successful, this project will lead to technologies for scaled manufacturing of low-cost, reduced rareearth-content, high-energy-density PMs.

Addressing the challenge of rare-earth and critical-materials-containing batteries, particularly in the emerging hybrid and electric vehicle transportation sectors, the Batteries for Electric Energy Storage in Transportation (BEEST) program invested $35 million in first-of-kind demonstration of new batteries and storage chemistries, structures and technologies. Disruptive technology approaches such as magnesium-ion and rechargeable metal-air batteries from earth-abundant resources are being investigated in this high-technology-risk/high-impact program. If successfully demonstrated through the BEEST program, these technologies will point the way towards batteries for transportation that will exceed the capabilities of the best state-of-the-art lithium-ion technologies.

Office of Energy Efficiency and Renewable Energy

The Office of Energy Efficiency and Renewable Energy (EERE)is supporting an applied magnet research project valued at $2 million (FY 2010) at Ames Laboratory. This project is focused on fabricating high-performance, cost-effective PMs that can be used for traction motors with an internal PM rotor design. Improving the alloy design and processing of PMs is essential to meeting performance and cost goals for advanced automotive electric drive motors. Technical requirements that the fully developed PM materials must meet for vehicle applications drive the project, such as adequate magnetic flux and coercivity for operation at elevated temperatures (180°C–200°C).

Requirements for material mechanical properties also impact manufacturing and assembly costs and may significantly influence the total motor cost. The Ames Laboratory project is first developing anisotropic magnets based on the high-temperature, rare-earth-based alloy previously designed for isotropic bonded magnets. As this alloy is rare-earth-based, it has the potential to address short- term needs but not long-term market concerns. The project is also developing high-performance magnet materials that do not contain rare earth constituents.

In addition to the magnet material research, EERE's Vehicle Technologies Program supports two projects valued at a total of $1.4 million (FY2010) at Oak Ridge National Laboratory investigating alternative motor designs that do not use rare earth PMs. Engineers are designing switched reluctance motors without PMs that can be manufactured with reduced fabrication costs and torque and speed characteristics similar to permanent magnet machines. A second project is developing a flux coupling motor with performance comparable to a PM machine. The objective is to produce cheaper traction drives and reduce system costs through lower current-rated components in the inverter and reduced transmission costs. In addition, in 2009, the Vehicle Technologies Program awarded $9.5 million to Toxco, to expand an existing battery recycling facility in Ohio and become the first U.S. facility to recycle lithium-ion vehicle batteries.

For wind power applications, reducing magnet size by developing higher flux density magnets is more important than consistent properties at elevated temperatures. EERE's Wind and Water Technologies Program is supporting QM Power, Inc., with $398,005 to develop a higher flux density PM generator. There are also much larger investments within EERE in battery, PV and lighting R&D that have key materials use implications.

4.3. Financial Instruments

DOE administers several programs that provide financial support for clean energy deployment. These include programs that provide loan guarantees and tax credits. None of these programs authorize DOE to provide financial support for mineral extraction or materials processing. However, several of these programs authorize DOE to support domestic manufacturing of component technologies (such as permanent magnets) that use critical materials.

Loan Guarantee Program

The Loan Guarantee Program (LGP) was established under Title XVII of the Energy Policy Act (EPAct) of 2005. It supports the production of clean energy components and end-use technologies. Section 1703 of EPAct 2005 authorizes loan guarantees supporting "new or significantly improved technologies to avoid, reduce or sequester air pollutants or anthropogenic emissions of greenhouse gases." The American Recovery and Reinvestment Act of 2009 (ARRA) added section 1705, which establishes additional loan guarantee authority to support "renewable energy systems, including incremental hydropower, that generate electricity or thermal energy, and facilities that manufacture related components."

The LGP lacks legal authority to provide loan guarantees for mineral extraction or processing because such projects do not meet the statutory requirements of either Section 1703 or 1705.

The LGP is authorized to provide loan guarantees to support domestic manufacturing of component technologies that use critical materials if those technologies meet the statutory tests. Projects supported by the program have the potential to affect market demand for key materials. For example, the LGP has recently issued loan guarantees to Solyndra ($535 million), Kahuku Wind Power ($117 million) and Beacon (Flywheel) ($43 million). Solyndra manufactures CIGS PV cells. Kahuku Wind and Beacon each use rare earth PMs.

Advanced Technology Vehicles Manufacturing Incentive Program

The Advanced Technology Vehicles Manufacturing (ATVM) Loan Program provides loans to automobile and automobile part manufacturers to re-equip, expand or establish manufacturing facilities in the United States to produce advanced technology vehicles or qualified components, and for the associated engineering integration costs. Vehicles with efficiency standards that will contribute to a clean energy economy are included in the definition of advanced technology vehicles. The ATVM lacks authority to directly support extraction and production of key materials. However, the ATVM issued loans to companies for projects that may affect the market demand of nickel metal hydride (NiMH) or Lithium ion batteries and NdFeB permanent magnet motors. These companies include Ford Motor Company ($5.9 billion), Nissan North America ($1.6 billion), Tesla Motors ($465 million) and Fisker Automotive ($529 million).

Tax Credits

The ARRA authorizes the Secretary of the Treasury, in consultation with the Secretary of Energy, to award tax credits for qualified investments in new, expanded or re-equipped domestic manufacturing facilities for clean energy technologies. The goal of the Advanced Energy Manufacturing Tax Credit—codified in Section 48c of the Internal Revenue Code—is to expand the domestic manufacturing industry for clean energy. Tax credits have been issued to manufacturers in a number of relevant energy technology areas, including solar thin film, LED lighting, wind turbine components and electric vehicles.

5. Other U.S. Government Programs

There is significant ongoing work in other federal agencies that supports and complements the U.S. Department of Energy (DOE) engagement on critical materials. Important contributions include collection and publication of data, analyses of demand, development of trade policies and support of research. In addition, there is a growing opportunity to coordinate and integrate relevant work through the White House Office of Science and Technology Policy (OSTP).

5.1. Office of Science and Technology Policy

Interagency collaboration on materials research and related policy is led by the OSTP. Since early 2010, OSTP has hosted an Interagency Working Group on Rare Earth Elements. This working group has coordinated interagency analysis and policy development relating to the evolving rare earth situation. In late 2010, a charter was signed for a Subcommittee on Critical and Strategic Mineral Supply Chains, under the Committee on Environment, Natural Resources and Sustainability (CENRS). The purpose of the new Subcommittee is to advise and assist the CENRS on policies, procedures and plans relating to risk mitigation in the procurement and downstream processing of critical and strategic minerals. Functions of the Subcommittee include identifying critical and strategic minerals and identifying cross-agency research and development opportunities.

5.2. U.S. Geological Survey

The U.S. Geological Survey (USGS) collects, analyzes and disseminates information on the domestic and international supply of and demand for minerals and materials essential to the U.S. economy and national security. USGS also provides assessments of undiscovered mineral resources in the United States and around the world. Researchers and decision makers use this information to understand the factors underlying an adequate and dependable supply of minerals and materials. This information also illuminates costs and risks related to the environment, energy and economics. The USGS National Minerals Information Center publishes reports in the annual Mineral Commodity Summaries and Minerals Yearbook for the group of rare earths, platinum-group metals, lithium, tellurium, indium and other key materials.

The public and private sectors rely on USGS minerals information and assessments to better understand the use of materials and the ultimate disposition of materials in the economy. USGS minerals information also informs the efficient use of national resources. In addition, USGS minerals information is used to forecast future supply of and demand for minerals. Domestic and international minerals information is used to analyze policies, formulate plans to deal with shortages and interruptions in mineral supplies and develop strategies to maintain a competitive position in the global economy.

5.3. U.S. Department of Defense

Studies and Analysis

Recognizing the evolution of the market for rare earth elements (REEs), in the summer of 2009 the Office of Industrial Policy/AT&L, Department of Defense (DoD) self-initiated a review of the U.S. supply chain. The study is based on available forecasts and data from multiple sources and as a result, most of the data are available only at the aggregate level of all REE. The study reviews the U.S. supply chain for both commercial and defense demand of REE. The study also assesses gaps in the supply chain and their potential implications for the Department.

The rationale for this effort included the U.S. dependence on a sole supplier that is not domestic, the importance of REE in certain defense applications and forecasts for a surge in demand for commercial end uses that could strain global supplies. Recent events in the global market for REE have reinforced the Department's concern regarding reliable and secure supplies of REE.

National Defense Stockpile Program

In the United States, stockpiling is largely the province of DoD, which maintains a National Defense Stockpile (NDS) managed by Defense Logistics Agency (DLA). The NDS Program was established in 1939 to preclude dependence on foreign sources in times of national emergency. NDS holdings grew to 90 commodities in 85 locations by 1994, when Congress first authorized the sale of excess NDS inventory. Since then, the NDS has liquidated approximately $7 billion in commodities and reduced its holdings to 25 managed commodities at 17 locations. However, the current NDS system focuses solely on the physical stockpiling of raw materials rather than the entire defense supply chain. It also requires separate authorizing legislation for each material.

At the direction of Congress in 2006, DoD initiated a review of the NDS led by the Office of the Secretary of Defense (OSD). The results of that review, presented in an April 2009 report to Congress, included a plan for establishing a comprehensive Strategic Materials Security Management System (SMSMS) that would identify, on an ongoing basis, those strategic and critical materials required for national security (OSD 2009).

The Strategic Military Stockpile Program (SMSP) concept would include limited physical stockpiles used in conjunction with friendly nation agreements and long-term supply chain partnerships to provide assurances for military equipment manufacturers regarding material price and availability. The OSD report also recommended holding physical reserves of 13 materials (including cobalt) while continuing to monitor and study 40 other materials (including gallium, indium, tellurium and yttrium).

5.4. Other Agencies and Departments

A number of other agencies and departments have important roles and interests in the global materials supply chain and related innovation system. Mine permitting is handled by the U.S. Department of the Interior and the U.S. Environmental Protection Agency (EPA). Global trade analysis and policy is under the U.S. Department of Commerce, the U.S. Department of State and the U.S. Trade Representative. The U.S. Department of State embassy officials report on relevant policies of host governments, as well as on private sector efforts and local markets. The National Institute of Standards and Technology (NIST) sets engineering standards for the manufacturing sector. In addition to DOE, the National Science Foundation, the National Oceanic and Atmospheric Administration, NIST and the EPA support relevant research.

6. Materials Strategies from Other Nations

Raw materials policies vary greatly between nations, due to differences in natural resources, systems of governance and industrial make-up. Different methods of addressing materials requirements provide meaningful lessons and may help inform DOE's approach to this issue. The raw materials policies of Japan, the European Union (EU), the Netherlands, China, South Korea, Australia and Canada represent a broad range of national interests, resource characteristics and policy goals. Table 6-1 outlines each nation's policy goals, business policy, research policy and materials of interest.

6.1 Japan

Japan's materials policy is based on the nation's limited domestic resources and the importance of many rare metals to the manufacturing of electronics and automobiles. The policy's goals, as outlined in the 2009 "Strategy for Ensuring Stable Supplies of Rare Metals," include (i) maintaining a stable supply of metals for Japanese industries by securing overseas sources of critical materials; (ii) recycling rare scrap metals; (iii) developing alternative materials; and (iv) stockpiling some rare metals (METI 2009). Japan's raw materials policy is guided by the Ministry of Economy, Trade and Industry (METI) and implemented by the Japan Oil, Gas and Metals National Corporation (JOGMEC) and the Japan Bank of International Cooperation (JBIC), with support from other ministries and government institutions. In 2007, the budget for Japanese mineral resource policy was roughly $70 million (Kawamoto 2008).

Table 6-1. Policy Goals, Business Policies, Research and Development Policies and Materials of Interest for Each Nation

Nation	Goal	Business Policy	R&D Policy	Materials of Interest
Japan	Secure a stable supply of raw materials for Japanese industries	• Funding for international mineral exploration • Loan guarantees for high-risk mineral projects • Stockpiling • Information gathering	• Substitution research funded through METI and MEXT • Exploration, excavation, refining and safety research funded through JOGMEC	Ni, Mn, Co, W, Mo, V**
European Union	Limit the impact of potential material supply shortages on the European economy	• Mineral trade policy for open international markets* • Information gathering* • Land permit streamlining* • Increased recycling regulations*	• Increased material efficiency in applications • Identification of material substitutes • Improve end-of-life product collection and recycling processes	Sb, Be, Co, Ga, Ge, In, Mg, Nb, REEs, Ta, W, Fluorspar and Graphite
Netherlands	Reduce material consumption to prevent global shortages by employing "managed austerity"	• Government-industry collaboration on material policy through the M2i Institute	• Substitutes of abundant or renewable materials • Processes for recycling depleting materials • Study consumption patterns as a result of policy	Ag, As, Au, Be, Bi, Cd, Co, Ga, Ge, Hg, In, Li, Mo, Nb, Nd, Ni, Pb, Pd, PGMs, REEs, Re, Ru, Sb, Sc, Se, Sn, Sr, Ta, Te, Ti, V, W, Y, Zn, Zr
China	Maintain a stable supply of raw materials for domestic use through industry consolidation, mitigating overproduction and reducing illegal trade	• Taxes and quotas on REE exports • Prohibition of foreign companies in REE mining • Industry consolidation • Unified pricing mechanisms* • Production quotas • Moratorium on new mining permits until mid- 2011	• Rare earth separation techniques and exploration of new rare earth functional materials • Rare earth metallurgy; optical, electrical, and magnetic properties of rare earths; basic chemical sciences of rare earths	Sb, Sn, W, Fe, Hg, Al, Zn, V, Mo, REEs
South Korea	Ensure a reliable supply of materials critical to Korean mainstay industries	• Financial support for Korean firms at overseas mines • Free Trade Agreements and MOUs with resource- rich nations • Stockpiling	• Recycling end-use products • Designing for recyclability • Substitute materials • Production efficiency	As, Ti, Co, In, Mo, Mn, Ta, Ga, V, W, Li and REEs
Australia	Maintain investment in the mining industry while fairly taxing the depletion of national resources	• Low tax on the value of extracted resources • High tax on mine profits • Tax rebates for mineral exploration • Fast turnaround for land permit applications	• Promote sustainable development practices in mining	Ta, No, V, Li and REEs
Canada	Promote sustainable development and use of mineral and metal resources, protect the environment and public health and ensure an attractive investment climate	• Promote a recycling industry and incorporate recycling as part of product design • Require accountability in environmental performance and mineral stewardship • Use life-cycle-based approach to mineral management and use	• Provide comprehensive geosciences information infrastructure • Promote technological innovation in mining processes • Develop value-added mineral and metal products	Al, Ag, Au, Fe, Ni, Cu, Pb, Mo

*proposed policy **current reserves.

JOGMEC is an independent administrative institution owned by the Japanese government that enacts government policy but independently controls its own budget and management. JOGMEC promotes a stable supply of metal resources through five activity areas:

- Providing partial funding for overseas field surveys through the Joint Basic Exploration Scheme
- Providing loan guarantees and other financial assistance to high-risk mine development projects
- Maintaining stockpiles of seven metals—nickel, chromium, manganese, cobalt, tungsten, molybdenum and vanadium—while closely monitoring the availability of Indium, rare earth elements, platinum, gallium, niobium, tantalum and strontium
- Gathering and disseminating information on mineral availability and policies in various nations
- Funding and engaging in scientific research on new types of exploration, mining and recycling (JOGMEC 2007).

According to a 2008 METI strategy statement in response to geopolitical developments in global mineral supply, the Japanese government will also provide diplomatic assistance to Japanese companies engaging in mining projects abroad by giving official development assistance to mining and transportation infrastructure projects (METI 2008).

Japanese firms are actively securing the raw materials needed for their operations. Toyota Motor Corporation established a rare earth task force to monitor its supply chain and, through its trading company Toyota Tsusho, invested in a rare earth mining joint venture in Vietnam in 2008 to export rare earths to Japan (AP 2010). Likewise, Japanese trading house Sumitomo Corporation established a joint venture in Kazakhstan with the goal of producing 3,000 tons of rare earths per year (Japan Looks Past 2010).

METI, the Ministry of Education, Culture, Sports, Science and Technology (MEXT) and the government-affiliated New Energy and Industrial Technology Development Organization (NEDO) also directly fund research projects on substitutes for and efficient use of rare metals. Recent research projects have focused on reducing the material used in rare metals technologies and substituting rare metals with more abundant ones. In October 2010, NEDO and Hokkaido University announced the development of a motor for hybrid and electric vehicles that does not use rare earth elements, instead utilizing magnets from less expensive and more common ferrite materials (Japan Looks Past 2010, Tabuchi 2010).

6.2. European Union

European Union nations rely heavily on imported rare metals and products containing rare metals. In response to recent demand increases caused by emerging technologies (e.g., tantalum use in cell phones), the European Commission established the Raw Materials Initiative to limit the impact that material supply shortages may have on the European economy.

The Raw Materials Initiative contains three policy pillars:

Maintain access to raw materials in world markets on the same conditions as international competitors—enforce World Trade Organization trade policy and the provision of development aid in resource-rich nations to support good governance, a sound investment environment and environmentally safe practices.

Establish EU framework conditions that foster a sustainable domestic supply of raw materials— maintain congruent data on mineral availability and mining regulations among member states; streamline the land permitting process for mining; support research on extraction and processing; and initiate university education programs for mining science.

Increase resource efficiency and recycling to reduce consumption of raw materials (Commission of the European Communities 2008)—fund research in reduced-material product designs, recycling and material substitutes; improve end-of-life product collection in all member states; and enforce export restrictions for recyclable waste.

The European Commission recently released a study assessing materials for criticality, as defined by the value each material adds to the European economy and the material's potential for an international supply shortage. Fourteen of the 41 materials studied were identified as exhibiting a high supply risk and high economic importance: antimony, beryllium, cobalt, fluorspar, gallium, germanium, graphite, indium, magnesium, niobium, platinum group metals, rare earth metals, tantalum and tungsten (European Commission, Enterprise, and Industry 2010).

The European Commission plans to solicit research proposals for deep sea mining, material substitutes and recycling and recovery of critical materials. Total research funding will be at least $34.5 million.

6.3. Netherlands

The Netherlands, a member of the EU, is developing its own rare metals strategy based on its concern that the rapid depletion of raw materials is partially due to over-consumption, and thus "managed austerity" should be part of the remedy. According to a report published by M2i – a public-private partnership between the Dutch government, universities and industry—the fruitful exploration for, and extraction of, rare metals will not continue to fulfill all of this century's needs. Instead, governments must prepare for material scarcity by promoting the substitution of plentiful or renewable materials for rare metals, more efficient use of depleting metals and efficient and productive recycling (M2i 2009). The rare metals strategy will be developed in collaboration between government, universities, industry and research organizations.

The Netherlands will use research and development funding to develop and implement these strategies. The Netherlands Organization for Scientific Research (NWO) established a research theme titled "Materials: Solutions for Scarcity", which includes the development of substitute materials (both mineral and bio-based) for depleting resources, processes for recycling metals in post-consumer products and social science research in policy effects on consumption patterns. As an initial budget, the organization has allocated $10 million over the next four years for the research theme.

6.4. China

As a major producer and rapidly increasing consumer of raw materials, China's policy goals are to maintain a stable supply of materials for the Chinese economy and reduce illegal mining, overproduction and smuggling of its domestic resources. These goals specifically concern REEs, of which China is the world's leading producer.

In the past decade, China has moved toward supporting domestic markets for REEs. In 2007, the Ministry of Commerce declared most rare earth elements and products to be strategic commodities and the State Council placed new restrictions on foreign investment in the REE sector. These developments prohibited outright foreign investment in REE mining—requiring that foreign investors form joint ventures with domestic firms in the processing of rare earth ores—and officially encouraged foreign investment in the more value-added manufacturing of rare earth magnets, metal alloys and powders. China further restricted foreign access by abolishing export tax rebates in 2005 and introducing a new REE export tax in 2006. Export tax rates have since been raised and now range from 15%–25% for different elements. Between 2004 and 2009, the overall REE export quota was reduced by more than 20% from about 65,000 tons to about 50,000 tons rare earth oxides (REOs). In July 2010, China further reduced its export quota to 30,258 tons REO, a 40%decrease from 2009.

While attempts to exert more control over the rare earth industry have recently gained momentum, there remain a number of challenges. Illegal mining and smuggling, for instance, are major issues for Chinese mining policy. In 2010, the Ministry of Land and Resources, responsible for issuing mining licenses, decided to stop issuing new licenses until mid-2011 (Muyuan 2010). However, illegal mining of REEs continues. Moreover, the enforcement of environmental and other mining regulations varies by province, which can lead to severe environmental degradation associated with rare earth mines. Poor environmental protection compliance across the industry often results in thorium residues being disposed into unlined tailing facilities and insufficiently treated water reaching nearby rivers.

To further protect domestic resources Chinese officials are planning to create sizable stockpiles as well. In February 2010, the regional government of Inner Mongolia authorized a "strategic reserve" of REEs to be established in the autonomous region (Yan and Yijun 2010). In October 2010, it was reported that Bautou Steel's plan to acquire and set aside up to 300,000 tons of rare earths within five years was approved by the Chinese government (China May Launch Rare 2010).

China has supported rare earths R&D efforts since the 1950s and currently sponsors two key national research programs and four state laboratories. Researchers focus on REE separation techniques; the exploration of new REE functional materials; and optical, electrical and magnetic properties of REEs. Other programs focus on REE basic chemical sciences including solid state chemistry, bioinorganic chemistry, chemical biology and separation chemistry. The Baotou Research Institute, established in 1963, focuses specifically on rare earth metallurgy, environmental protection, new materials and applications in traditional industries. The Chinese Society of Rare Earths publishes two academic journals dedicated to rare earths: the *Journal of Rare Earth* and the *China Rare Earth Information (CREI) Journal*.

6.5. Republic of Korea

Like Japan, the Republic of Korea (South Korea) focuses its materials strategy on minerals deemed critical to the competitiveness of its commercial industries in consumer electronics, information technology, automobile manufacturing and clean energy. A policy plan, "Plans for Stable Procurement of Rare Metals," is being drafted by the Ministry of Knowledge Economy and seeks to spend $15 million by 2016 in order to secure 1,200 metric tons of rare earth elements in addition to developing domestic mines for other rare metals (AP 2010). South Korea has identified 56 elements of interest, including 11 that it defines as "strategic critical," based on rarity, unfavorable geological distribution and price instability. South Korea is seeking to decrease its heavy dependence on imported raw material inputs through the following four-pronged approach:

- Government-backed investment in the exploration of foreign sources of rare metals
- Increased stockpiles with the flexibility to meet the country's needs
- Reduced consumption through the development of substitutes
- Increased recycling and reuse of materials from end-use products (Bae 2010)
- South Korea's base metals stockpile program—modeled after Japan's—has announced plans to expand its holdings to 15 metals, including cobalt and titanium (OSD 2009).

South Korean firms are also beginning to implement this strategy with the help of the government. The state-owned Korea Resources Corporation (Kores) plans to spend $285.2 million in 2010 to develop overseas mines of lithium, nickel, uranium, copper and manganese in Africa and Latin America (Ha-won 2010). In March 2010, Kores partnered with Korean steelmaker Posco to take a controlling stake in China's Yongxin Rare Earth Metal Co., enabling Posco to bypass China's export quotas of rare earth oxides by achieving direct access and gaining the ability to export rare earths back to South Korea legally (Yang 2010).

South Korea plans to focus its R&D efforts on 40 core technologies through the Korea Institute of Industrial Technology, a government-funded research center (Han 2010). Recycling rare metals from end-use products and designing for recyclability at the production stage are particularly important. Scrap piles of used products (called "urban mines") can be used to recover rare metals and, based on the standardization of recycling systems, can be placed at the beginning of the supply chain as additional resource inputs.

6.6. Australia

Mining accounts for 7% of Australia's national economy (USGS 2008). Australia's national mining policy is managed by the Department of Resources, Energy, and Tourism. Policy goals involve balancing a stable investment environment that promotes mining industries, fair regulation and taxation of national resources with sustainable extraction and use of finite earth materials. Major issues in mining policy include taxes, permitting, information gathering and the stockpiling of mineral reserves.

Australia imposes taxes on its mining sector principally under state and territory jurisdictions, although the Australian federal government has suggested it may establish a profit tax on certain mineral commodities at a rate of 30%–40% (Smith 2010). In order to promote exploration and stabilize investments, Australia allows mining companies to deduct expenses or claim rebates for exploration costs and to roll over losses or profits between years. Australia is consistently ranked the country with the fastest permitting time by the international mining consultant firm Behre Dolbear (Behre Dolbear 2010).

6.7. Canada

Canada is the world's largest exporter of minerals and metals, with natural resource mining accounting for 4% of its gross domestic product. National mining policy is managed by Natural Resources Canada, but primary responsibility for mining oversight falls under provincial jurisdiction. At the federal level, the government uses a mix of policies in finance and taxation, regulatory efficiency and investment and export promotion to maintain a globally competitive industry (Natural Resources Canada 1996). Canada also maintains a relatively flexible and favorable regulatory regime that seeks to avoid duplication, minimize uncertainty and delays and harmonize federal and provincial rules. While Canada has extensive mining and environmental regulations, it still ranks ninth out of 25 nations in terms of permitting time by Behre Dolbear (Behre Dolbear 2010). Canada stores copper, gold, lead, molybdenum, nickel, silver and zinc in quantities from 0.5%–4% of national annual production levels.

7. Supply and Demand Projections

In this chapter, we explore the extent to which more widespread deployment of clean energy technologies could lead to imbalances of supply and demand for rare earth elements (REEs) and other key materials. To assess these risks, we compare the projected levels of demand for each key material with projected levels of supply.

7.1. General Methodology for Estimating Future Demand for Key Materials

Future demand for key materials will come from both clean energy and non-clean energy sources. This section discusses the general methodology for estimating future demand for key materials in both categories.

The first step in projecting demand for key materials is to estimate expected demand in non-clean-energy technologies. These include mobile communication devices, polishing powders and flat screen televisions (discussed in more detail later in Chapter 3 and Appendix A). Time and resource constraints precluded projections for each of the many non-clean energy technologies that use REEs and other key materials. Instead, this analysis assumes that demand for key materials in non-clean energy technologies increases at the rate of growth for the global economy projected in the International Energy Agency's *World Energy Outlook*

2009. [73] Accordingly, that demand is projected to increase from its 2010 level at a compound annual growth rate of 3.3% during the period from 2010–2015, and at a compound annual growth rate of 3% from 2015–2025. To estimate non-clean energy demand in 2010 for each material, an estimate of current clean energy demand (Trajectory B as described below) is subtracted from total material demand in 2010.

The next step in projecting demand for key materials is to estimate expected annual demand in clean energy technologies. This analysis focused on four main components of rare earth elements and other key materials among clean energy technologies. Those four components are:

- *Permanent magnets* (used in wind turbines and electric vehicles)[74]
- *Advanced batteries* (used in electric vehicles)
- *Thin-film semiconductors* (used in photovoltaic (PV) power systems)
- *Phosphors* (used in high-efficiency lighting systems)

Estimates of future demand for key materials in clean energy applications were calculated as the product of three factors:

1. **Deployment:** total units of the generic clean energy technology in a given year
2. **Market Share:** the percentage of installations captured by a specific clean energy technology
3. **Material intensity:** demand for the material in each unit of the clean energy component

Looking out over the period from 2010–2025, the rate of future technology deployment for wind turbines, advanced vehicles, photovoltaic power systems and high-efficiency lighting is highly uncertain. Also uncertain are the particular components that will succeed and support technology deployment. To assess the risk of future material supply-demand imbalances, a *High Penetration case* and a *Low Penetration* case were developed. The *High Penetration* case combines a **high level of global deployment** for the generic technology with a **high market share** captured by the specific clean energy technology. The *Low Penetration* case combines a **low level of global deployment** for the generic technology with a **low market share** captured by the specific clean energy technology.

There is also significant uncertainty about the amount of material needed for each clean energy application, **the material intensity**, looking forward to 2025. To account for this uncertainty, a *Low Materials Intensity* case was constructed reflecting a low, but feasible estimate of material required per unit of technology deployed. A *High Materials Intensity* case was similarly constructed describing a high but feasible estimate of material required.

High and low values for material intensity and market share represent best estimates of these parameters in the short and medium term. Global deployment rates, on the other hand, are not intended to be predictive. For magnets, batteries and photovoltaics, global deployment is based on International Energy Agency (IEA) estimates of technology deployment under different assumptions about national policies and clean energy objectives. For phosphors, global deployment is based on linear growth of historical demand.

For each material, the high and low assumptions for rates of technology deployment, market share and material intensity were combined and added to the non-clean energy demand to develop four distinct demand trajectories. Two trajectories, labeled **Trajectory A** and **Trajectory B**, reflect the slower rate of penetration for each application and represent combinations of the *Low Penetration* case with the high and low assumptions respectively for material intensity. Similarly, two trajectories, called **Trajectory C** and **Trajectory D**, represent combinations of the *High Penetration* case with the respective low and high material intensity assumptions. In the figures comparing future supply and demand for each material that appear later in this chapter, Trajectories A and B are shown in blue and Trajectories C and D are shown in green. Also in those figures, high material intensity cases (Trajectories B and D) are shown as solid lines and low material intensity cases (Trajectories A and C) are shown as dashed lines. The characteristics describing Trajectories A, B, C and D are in Table 7-1.

None of the four trajectories analyzed in this report is intended to imply a prediction of future demand for clean energy technologies or key materials used in making them. That demand will depend on a number of factors, including technological progress, policy consistency and macroeconomic conditions. Instead, the trajectories are intended to illustrate a range of future possibilities and explore the impact of different assumptions concerning technology deployment rates, market shares and material intensity on future requirements for rare earth elements and other key materials. Trajectories A and D will represent the lower and upper extremes, respectively, for probable material demand.

In the following sections, these four trajectories are calculated for each key material in clean energy technologies. For key materials that are used in several clean energy technologies, the trajectories of future demand are presented as an aggregate for all relevant technologies. The contribution of each application is noted in the discussion of the figure. For example, Trajectory A demand for neodymium represents the sum of non-clean energy demand plus the Trajectory A requirements for magnets used in wind turbines plus the Trajectory A requirements for magnets used in vehicles with electric drive-trains.

7.2. Short- and Medium-Term Supplies of Key Materials

The Earth's crust contains sufficient rare earth elements and other key materials to meet projected demand in the decades ahead. However, short- to medium-term growth demand for these materials in clean energy technology and other applications may strain the ability of supply chains to provide global markets with a smooth flow of materials at stable prices. The first step in the supply chain— mines—often requires a number of years to be brought on line. This section describes potential production increases of rare earth elements and other elements in the next five years and longer. The figures in sections 7.3–7.6 show 2010 production with a solid red line and total production by 2015 with a dashed red line.

Short- and Medium-Term Supplies of Rare Earth Elements

Around the world, there are many promising mineral deposits that could be developed to meet future growth in demand for rare earths. These deposits are found on at least six continents, including significant resources in Asia, Australia, North America and Africa. Whether a deposit can be mined economically will depend on a number of factors, including rare earth prices, regulatory requirements and improvements in extraction and separation technologies.

Table 7-1. Assumptions to Estimate Future Trajectories of Material Demand

Trajectory of Demand	Market Penetration		Material intensity of the Clean Energy Component
	Global Deployment Level of the Generic Technology	*Market Share of Specific Clean Energy Technology*	
Trajectory A	Low	Low	Low
Trajectory B	Low	Low	High
Trajectory C	High	High	Low
Trajectory D	High	High	High

Table 7-2. Current and Projected Future Rare Earth Supply by Element[75]

Rare Earth Supply by Element: Production Sources and Volume (tonnes/yr)										
	Estimated 2010 Production	Assumed Additional Production by 2015							Total Additional Production by 2015	Estimated 2015 Production
		Mountain Pass (USA)	Mt. Weld (Australia)	Nolans Bore (Australia)	Nechalaco (Canada)	Dong Pao (Vietnam)	Hoidas Lake (Canada)	Dubbo Zirconia (Australia)		
Lanthanum	33,887	6,640	3,840	2,000	845	1,620	594	585	16,124	50,011
Cerium	49,935	9,820	6,855	4,820	2,070	2,520	1,368	1,101	28,554	78,489
Praseodymium	6,292	860	810	590	240	200	174	120	2,994	9,286
Neodymium	21,307	2,400	2,790	2,150	935	535	657	423	9,890	31,197
Samarium	2,666	160	360	240	175	45	87	75	1,142	3,808
Europium	592	20	90	40	20		18	3	191	783
Gadolinium	2,257	40	150	100	145		39	63	537	2,794
Terbium	252		15	10	90		3	9	127	379
Dysprosium	1,377		30	30	35		12	60	167	1,544
Yttrium	8,750	20	60		370	35	39	474	998	9,748
TOTAL	**127,315**	19,960	15,000	9,980	4,925	4,955	2,991	2,913	**60,724**	**188,039**

Source: Industrial Minerals via Watts (2010).

Potential production from new mines coming on line before 2015 is shown in Table 7.2. The Mount Weld mine (Australia) is currently slated to come on line in 2011. (Ores from the Mount Weld mine will be sent to Malaysia for separation and refining.) The Mountain Pass mine (California) is projected to come on line in late 2012, after setting up new refining and manufacturing facilities at the site. Projected annual production from these two mines is also shown in the figures in sections 7.3–7.6. Additional mines with the potential to come on line before 2015 include Hoidas Lake (Canada), Dubbo Zirconia (Australia), Dong Pao (Vietnam), Nolans Bore (Australia) and Nechalacho (Canada). Figure 7-1 shows the deposits in Table 7-2, as well as a global distribution of 14 additional promising deposits that could be candidates for development in the medium term (Watts 2010).

Selected rare earth projects outside China

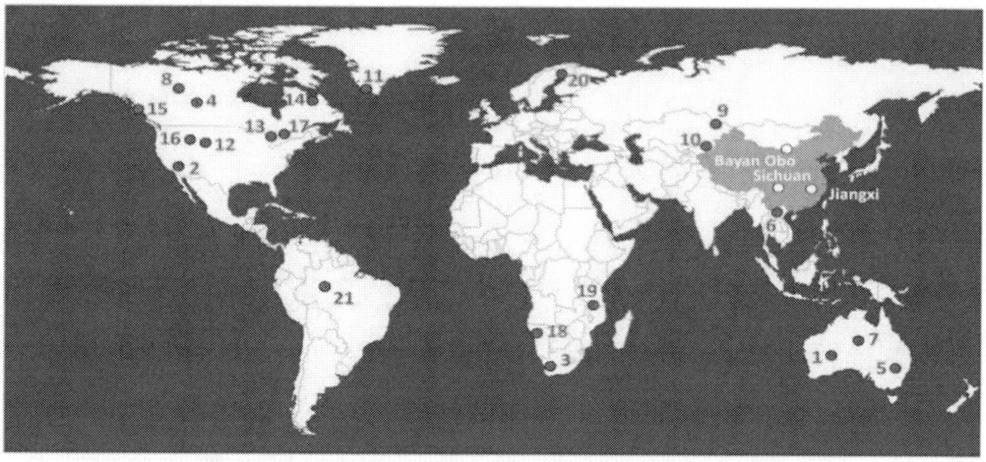

(1) Lynas Corp, (2) Molycorp Minerals, (3) (4) Great Western Minerals, (5) Alkane Resources, (6) Vietnamese govt/Toyota Tsusho/Sojitz, (7) Arafura Resources, (8) Avalon Rare Metals, (9) Kazatomprom/Sumitomo, (10) Stans Energy, (11) Greenland Minerals and Energy, (12) Rare Element Resources, (13) Pete Mountain Resources, (14) Quest Rare Minerals, (15) Ucore Uranium, (16) US Rare Earths, (17) Matamec Explorations, (18) Etruscan Resources, (19) Montero Mining, (20) Tasman Metals, (21) Neo Material Technologies/Mitsubishi

Sources: Kingsnorth (see footnote 76), Roskill (2010) and USGS (2010).

Figure 7-1. Current and projected rare earth projects.

Emerging Supplies of Other Key Materials

Additional elements playing key roles in the development of clean energy applications include indium, gallium, tellurium, cobalt and lithium. Table 7-3 describes current production and potential additional supply (based on current production capacity and estimated additional production capacity) through 2015 for each of these elements.

According to the Indium Corporation, one of the world's largest producers of indium, the 2010 global primary production of virgin indium is around 480 tonnes, while reclaimed indium contributes another 865 tonnes (Indium Corp. 2010). The estimated additional supply of indium coming on line by 2015 is based on a combination of assuming maximum capacity utilization of primary production (600 tonnes), an 80% scrap recovery from ITO processing (960 tonnes) and several new supplies of virgin indium brought on line by 2015.[76]

In 2010, only about 10% of alumina producers extract gallium as a byproduct of alumina processing. The remainder of producers find it too expensive to extract the gallium and thus treat gallium as an impurity in the aluminum refining process. The Indium Corporation estimated that in terms of global supply of gallium in 2010, about 125 tonnes of global primary production of gallium was supplemented by an additional 82 tonnes of globally reclaimed gallium (Indium Corp. 2010).

The estimated 2010 global tellurium production is based on the 2008 global primary production of tellurium reported by USGS due to the fact that the 2010 estimate was not yet available and the 2009 global primary production number reflects conditions when the recent economic recession had the most impact. Additional tellurium supply is based on the assumption that approximately 60% of the 1,200 tonnes/yr copper anode slimes based potential tellurium supply indicated by the USGS in the *2008 Tellurium Yearbook* will come online by 2015. Global tellurium production could grow by about a factor of four from increased levels of extraction from copper anode slimes by 2020, but may grow more slowly if copper refiners move away from the electrolytic process as the quality of copper ore declines. Additional supplies of tellurium also may be produced from tellurium mining projects and from efforts to recover tellurium from gold concentrates. However, no specific assumptions were made about these potential additional supplies due to insufficient information.

The estimated 2010 global primary cobalt production is based on the 2008 global primary production of cobalt reported by USGS, as using the production number from 2009, the last year of available data, would reflect conditions during the recent recession. There is a long list of potential additional sources of cobalt production. These include nickel and copper deposits in Canada, Western Australia, Democratic Republic of Congo (DRC), Zambia and Madagascar. In aggregate, these projects represent the potential to add up to 197,830 metric tons of cobalt per year to world supplies (USGS 2008d). Because of concerns about DRC's political and social stability, separate production values for Congo and non-Congo mines appear in the figures in sections 7.3–7.6.

Table 7-3. Current and Future Supply of Additional Elements Assessed

Supply of Other Elements Assessed: Production Sources and Volume (tonnes)				
	Estimated 2010 Production	**Potential Sources of Additional Production between 2010 and 2015**		**Estimated 2015 Supply**
		Additional amount	***Sources***	
2 Indium	1,345	267	Recovery (co-produced) from additional zinc production mainly and recycling	1,612
allium	207	118[77]	Recovery (co-produced) from additional alumina and bauxite production[78]	325
Tellurium	500	720	Recovery (co-produced) from copper anode Slimes	1,220
Cobalt	75,900	197,830	Mines	273,730
Lithium (carbonate equivalent)	134,600	115,400	Mines[79]	250,000

Sources: USGS 2008a-e and Evans 2010.

For lithium, the 2010 production number is estimated lithium carbonate equivalent (LCE) production available in 2010. In the near- to mid-term future, additional low-cost evaporative lithium resources may be developed from the high desert brines in Argentina and Chile, as well as from the geothermal brines in the Western United States.[80] Due to the uncertain nature of the lithium claims by emerging companies, this report takes into account only the expansion plans of the current lithium producers for the additional estimated supply by 2015.

Table 7-4. Assumptions for Key Materials in Magnet Technologies

	Technology	Assumption	Low Penetration	High Penetration
Deployment in 2025	Wind	Onshore Wind Turbine Additional Capacity (GW)	23.6	48.6
	Wind	Offshore Wind Turbines Additional Capacity (GW)	4.9	17.0
	Vehicles	Sales of Hybrid Electric Vehicles (HEVs) (millions)	4.2	19.1
	Vehicles	Sales of Plug-in Hybrid Electric Vehicles (PHEVs) (millions)	0.002	13.2
	Vehicles	Sales of All Electric Vehicles (AEVs) (millions)	0.001	4.6
Market Share	Wind	Onshore Wind Turbines using RE Magnets	10%	25%
	Wind	Offshore Wind Turbines using RE Magnets	10%	75%
	Vehicles	HEVs, PHEVs, and AEVs using RE Magnet Motors	100%	100%
Materials Intensity	Wind	Average Weight of Magnets per MW (kgs)	400	600
	Vehicles	Average Weight of Magnets per vehicle (kgs)	1.0	2.0
	Wind and Vehicles	% Weight of Magnets that is Neodymium	31%	31%
	Wind and Vehicles	% Weight of Magnets that is Dysprosium	5.5%	5.5%

7.3. Trajectories of Future Demand for Rare Earth Elements in Magnet Technologies

The use of rare earth permanent magnets (PMs) in vehicles and wind turbines is described in Chapter 2. These magnets incorporate several key materials, most notably neodymium and dysprosium.[81] The assumptions for deployment, market share and material intensity used to create trajectories of future demand for key materials in magnet technologies are presented in Table 7-4.

In 2010, neodymium-iron-boron (NdFeB) magnets dominate the market for high-efficiency traction motors in hybrid electric vehicles (HEVs). It is assumed that this trend continues and that NdFeB magnets are also used in all future plug-in hybrid electric vehicles (PHEVs) and electric vehicles (EVs). The market share for turbines with REE PMs is currently small, but projected to increase significantly over time as large capacity turbines using these motors enter the market in greater numbers.

Future deployments of onshore and offshore wind turbines were based on the International Energy Agency's *World Energy Outlook* (IEA WEO 2009). Low deployment is based on the "**2009 Reference Case**", as it represents a "baseline vision" of how energy markets are likely to evolve, taking into account only the array of policies and measures currently in place (but not necessarily fully- implemented) by mid-2009 (IEA 2009). High deployment of wind turbines is based on the "**450 Scenario**" in IEA WEO 2009, which identifies a set of technology deployment rates and technology assumptions that are capable of stabilizing atmospheric greenhouse gas concentrations at 450 parts per million by volume (ppmv) of CO2-equivalent by 2030.

Future deployment cases for vehicles with electric drive-trains (HEVs, PHEVs and EVs) were based on the International Energy Agency's 2010 *Energy Technology Perspectives* (IEA ETP 2010) with its detailed breakdown of annual deployment among various types of vehicles. The low deployment case is based on the "**2010 Baseline Case**," which like the IEA WEO "2009 Reference Case," assumes that governments introduce no new energy and climate policies after 2009. The high deployment case for light-duty vehicles with electric drive-trains is based on the IEA ETP 2010 "**Blue Map Scenario**." The IEA developed the "Blue Map Scenario" to illustrate a least-cost technology deployment scenario designed to reduce global, energy-related CO2 emissions by 50% from 2005 levels in 2050 (IEA 2010).

The development of the market share and material intensity assumptions is presented in detail in Appendix B.

Figure 7-2 contains projections of global requirements for neodymium oxide (Nd_2O_3) in all technologies, including wind turbines and vehicles with electric drive-trains, during the period 2010– 2025, as well as the 2010 and 2015 supply estimates. These amounts are given in terms of neodymium oxide, because this is the commercial feedstock from which the neodymium metal is refined and NdFeB magnets are fabricated. Also included in Figure 7-2 are supply estimates for 2010, 2010 plus two individual mines that are close to ramping up operations and an estimate for 2015 supply.

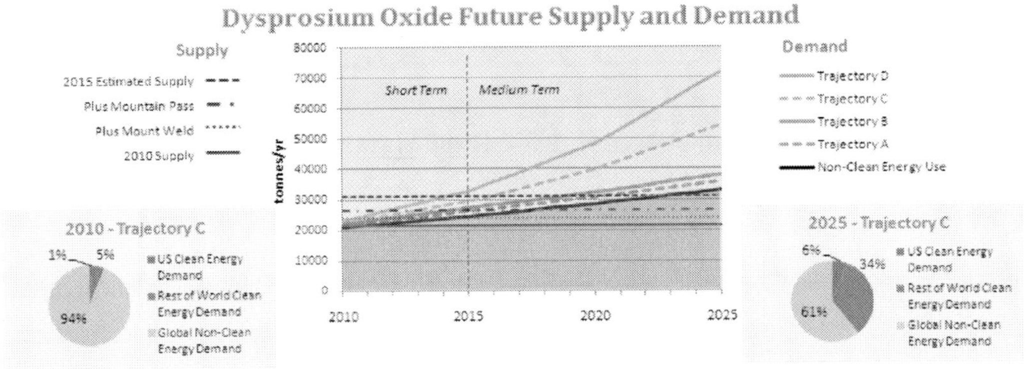

Figure 7-2. Future demand and supply for neodymium oxide.

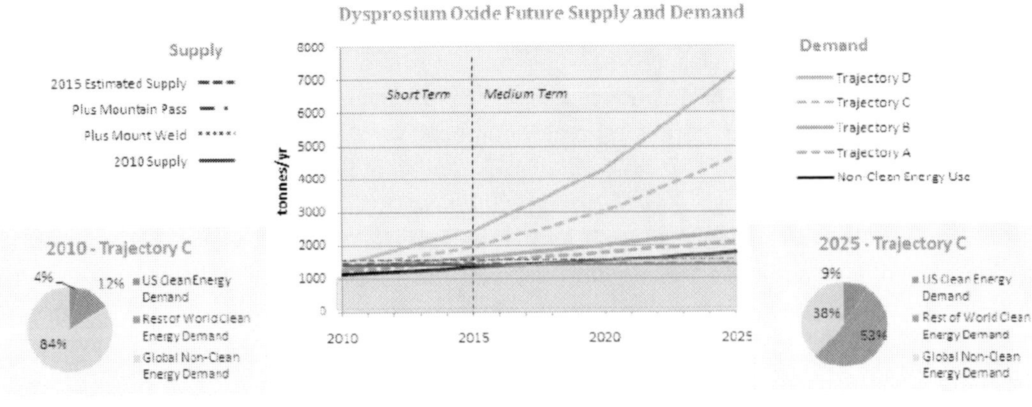

Figure 7-3. Future demand and supply for dysprosium oxide.

Figure 7-2 shows that the basic availability of neodymium oxide is adequate in the short term. Under high penetration scenarios (Trajectories C and D), clean energy represents a growing proportion of total neodymium oxide demand. Global demand for wind turbines and electric drive-train vehicles forms a significant percentage (40%) of total neodymium demand in 2025 under Trajectory C. Neodymium demand in vehicles contributes roughly five times higher demand than demand in wind turbines. Demand exceeds supply in 2015 only under Trajectory D. Projected non-clean energy demand alone will exceed projected 2015 supply before 2025. Deployment rate, not material intensity, is the biggest driver of demand, suggesting that R&D breakthroughs are needed to reduce neodymium content in magnets and batteries or development of substitutes.

Figure 7-3 illustrates the ranges of projections of global requirements for dysprosium oxide (Dy_2O_3) in magnet for wind turbines and vehicles, as well as non-clean energy use during the period 2010– 2025. These amounts are given in terms of dysprosium oxide because this is the commercial feedstock from which dysprosium metal is refined and NdFeB-AH magnets are fabricated. Also included in Figure 7-3 are supply estimates for 2010, 2010 plus additional individual mines and an estimate for 2015 supply.

Figure 7-3 shows that the basic availability of dysprosium oxide is tight in the short term. Electric drive-train vehicles represent roughly four times the demand for dysprosium oxide than wind turbines in 2025 under Trajectory C. Anticipated new mines will provide relatively little new supply, an additional 12% supply, by 2015. Global demand exceeds projected 2015 supply under all four trajectories in the beginning of the medium term. Global clean energy demand as a percentage of total demand for dysprosium increases dramatically from 16% in 2010 to 62% in 2025 under Trajectory C. The developing supply-demand imbalance in the medium term under all trajectories highlights the importance of R&D on alternative approaches to heat management (a main function of the dysprosium content) in magnets or substitutes for NdFeB magnets in general in clean energy technologies. Non-clean energy demand alone will lead to a supply-demand mismatch by the middle of the medium term under the assumed trajectory, highlighting the need for corresponding material intensity improvements or substitutes in non-clean energy technologies. Ceramic high-temperature superconductors may be a competitive substitute for NdFeB permanent magnets in wind turbines and this could lessen the demand for neodymium and dysprosium.

7.4. Trajectories of Future Demand for Rare Earth Elements in Battery Technologies

The use of nickel metal hydride (NiMH) and lithium-ion (Li-ion) batteries in vehicles with electric drive-trains is discussed in Chapter 2. These batteries incorporate a variety of key materials (e.g., lanthanum, cerium, neodymium, praseodymium, cobalt and lithium). For the purposes of this analysis, it is assumed that all future HEVs will use nickel metal hydride batteries, while PHEVs and EVs will rely on lithium-ion batteries. PHEVs are assumed to use a battery large enough to provide 40 miles of "electric-only" propulsion. EVs are assumed to use a battery large enough to provide 100 miles of propulsion on a single charge.

The assumptions for vehicle deployment, market share of the battery types and material intensity of the batteries used to create trajectories of future demand for key materials in battery technologies are presented in Table 7-5.

Table 7-5. Assumptions for Key Materials in Battery Technologies

	Technology	Assumption	Low Penetration	High Penetration
Deployment in 2025	Vehicles	Sales of Hybrid Electric Vehicles (HEVs) (millions)	4.2	19.1
	Vehicles	Sales of Plug-in Hybrid Electric Vehicles (PHEVs) (millions)	0.002	13.2
	Vehicles	Sales of All Electric Vehicles (AEVs) (millions)	0.001	4.6
Market Share	Vehicles	% PHEV-40s with Li-ion batteries	100%	100%
	Vehicles	% AEV-100s with Li-ion batteries	100%	100%
	Vehicles	% HEVs with NiMH batteries	100%	100%
	Technology	**Assumption**	**Low Intensity**	**High Intensity**
Materials Intensity	Vehicles	Average Weight of Lanthanum per NiMH battery (kg)	0.49	0.73
	Vehicles	Average Weight of Cerium per NiMH battery (kg)	0.69	1.03
	Vehicles	Average Weight of Neodymium per NiMH battery (kg)	0.20	0.31
	Vehicles	Average Weight of Cobalt per NiMH battery (kg)	0.44	0.66
	Vehicles	Average Weight of Cobalt per PHEV-40 Li-ion battery (kg)	0.00	3.77
	Vehicles	Average Weight of Cobalt per EV-100 Li- ion battery (kg)	0.00	9.41
	Vehicles	Average Weight of Lithium per PHEV-40 Li-ion battery (kg)	1.35	5.07
	Vehicles	Average Weight of Lithium per EV-100 Li- ion battery (kg)	3.38	12.68

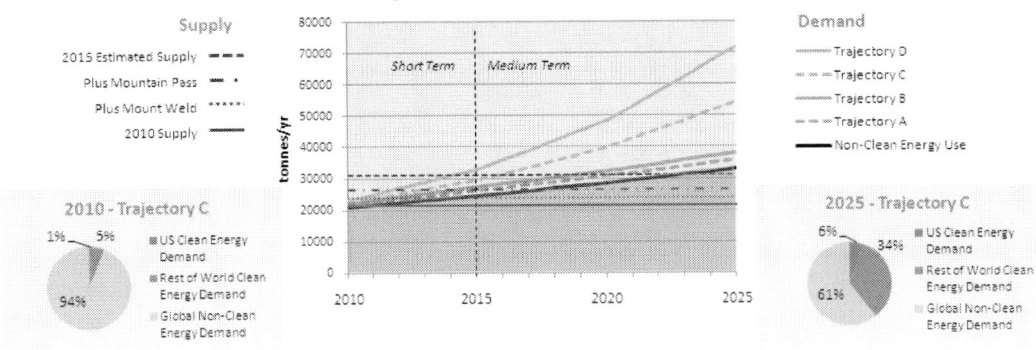

Figure 7-4. Future demand and supply for cobalt.

The International Energy Agency's *Energy Technology Perspectives* (IEA ETP 2010) "**2010 Baseline Case**" was selected as the basis for the *Low Penetration* case for light-duty vehicles with electric drive-trains. This "2010 Baseline Case" follows the outlines of the IEA WEO "2009 Reference Case," illustrates the total number of light-duty vehicles manufactured and sold each year and provides a detailed breakdown of annual deployment among various types of vehicles with electric drive-trains (IEA 2010). Like the IEA WEO "2009 Reference Case," the IEA ETP "2010 Baseline Case" assumes that governments introduce no new energy and climate policies after 2009.

The *High Penetration* case for light-duty vehicles with electric drive-trains is based on the IEA ETP 2010 "**Blue Map Scenario**." The IEA developed the "Blue Map Scenario" to illustrate a least-cost technology deployment scenario designed to reduce global, energy-related CO_2 emissions by 50% from 2005 levels in 2050 (IEA 2010). It contains projections of the total number of light-duty vehicles sold in each year and a breakdown of these vehicles by type of drive-train.

The development of the market share and material intensity assumptions is presented in detail in Appendix B.

Figures 7-4 through 7-7 illustrate the supply-demand picture in the future for cobalt, lithium, lanthanum and cerium. Figure 7-5 displays all trajectories, however, A and B overlap as vehicle deployment is so low in these cases. Figures 7-6 and 7-7 display supply and demand in terms of oxide because this is the commercial feedstock. Figures 7-6 and 7-7 also include future demand for lighting technologies, though this demand is far less than demand within vehicle batteries (less than 1,250 tonnes in 2025 under Trajectory D for both materials).

Figure 7-4 shows that the basic availability of cobalt appears more than adequate in the short to medium term, even where global clean energy demand increases dramatically under Trajectory D. Non-clean energy technologies represent the vast majority of cobalt global demand in all but Trajectory D. Additional supply capacity by 2015 appears to be more than sufficient to meet demand, even without mines from the Congo.

Figure 7-5 shows that the basic availability of lithium carbonate appears to be adequate in the short term. Global clean energy demand as a percentage of total demand increases dramatically from near zero to 49% in 2025 under Trajectory C. This increase is attributable to the rapid deployment of electric vehicles; 13 million plug-in hybrids and 4.6 million all-electric vehicles are sold in 2025 under Trajectory C. Global demand exceeds expected 2015 supply before 2025, but with the high levels of resources available, existing producers appear to be able to increase capacity beyond 2015 to meet global demand. To meet global lithium carbonate demand in 2025 under Trajectory C, an additional 100,000 tonnes per year of supply is needed over 2015 supply. Supply would need to more than triple over 2015 supply to meet demand for the high-clean-energy-penetration, high-materialintensity Trajectory D.

Figure 7-6 shows that the basic availability of lanthanum oxide is adequate in the short term. 2015 supply is able to meet demand under all trajectories until the middle of the medium term. Global demand for lanthanum in electric drive-train vehicle batteries as a percentage of total demand increases from 2% in 2010 to 19% in 2025 under Trajectory C. The clean energy demand trajectories also include the demand for lanthanum in lighting phosphors that is less than 1,000 tonnes per year in all years and trajectories (the assumptions behind lighting trajectories are described in more detail later in this chapter). If supply does not continue to increase after 2015, non-clean energy demand alone will exceed supply at the end of the medium term. Clean energy demand will exacerbate the supply-demand mismatch in the medium term without significant additional production, suggesting the importance of alternate battery technologies. To meet global lanthanum oxide demand in 2025 under Trajectory C, an additional 14,000 tonnes per year of supply is needed over 2015 estimated supply.

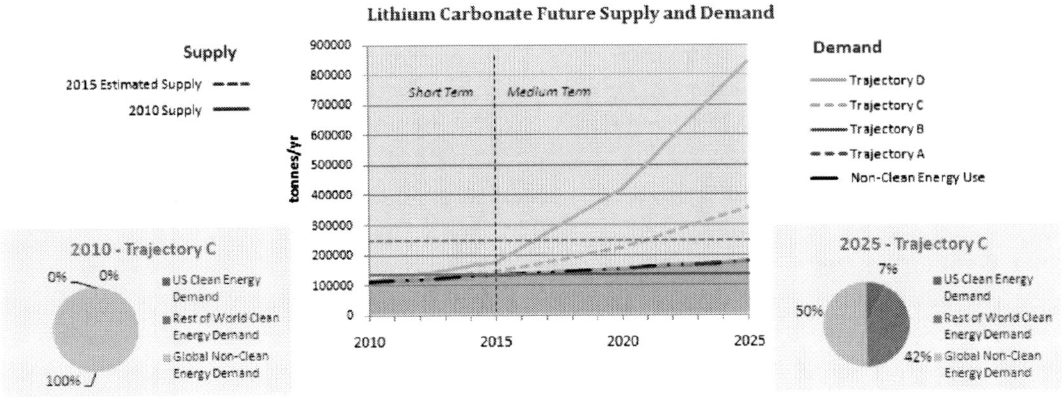

Figure 7-5. Future demand and supply for lithium carbonate.

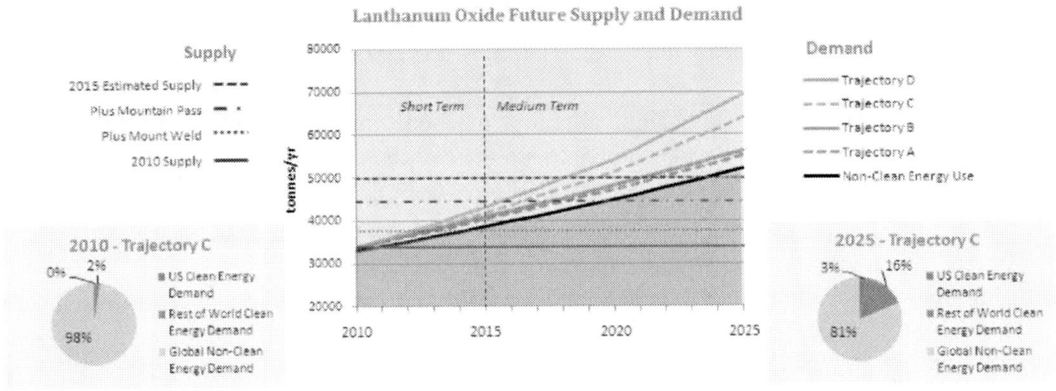

Figure 7-6. Future demand and supply for lanthanum oxide.

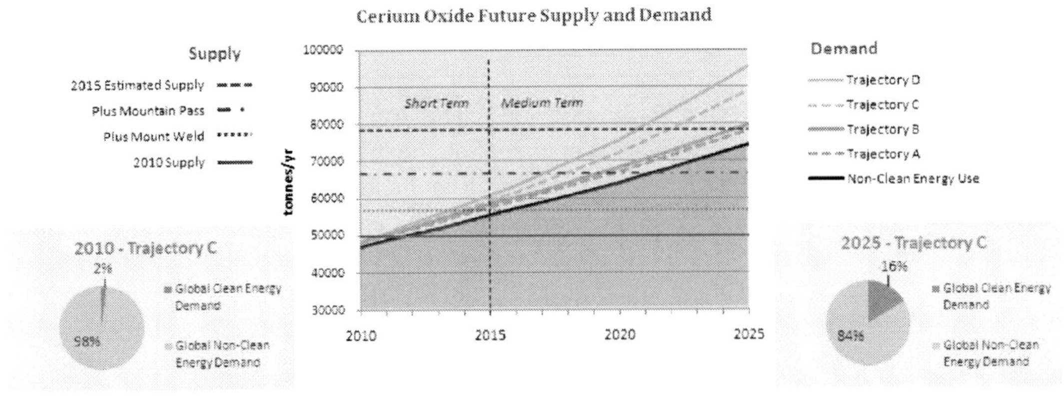

Figure 7-7. Future demand and supply for cerium oxide.

Table 7-6. Assumptions for Key Materials in PV Technologies

	Technology	Assumption	Low Penetration	High Penetration
Deployment in 2025	PV	Added Total PV Capacity (GW)	10.8	29.9
Market Share	PV	CIGS % of Added PV Capacity	10%	50%
	PV	CdTe % of Added PV Capacity	10%	50%
	Technology	**Assumption**	**Low Intensity**	**High Intensity**
Materials Intensity	PV	Avg Content of Indium per CIGS GW (tonnes)	16.5	110
	PV	Avg Content of Gallium per CIGS GW (tonnes)	4	20
	PV	Avg Content of Tellurium per CdTe GW (tonnes)	43	145

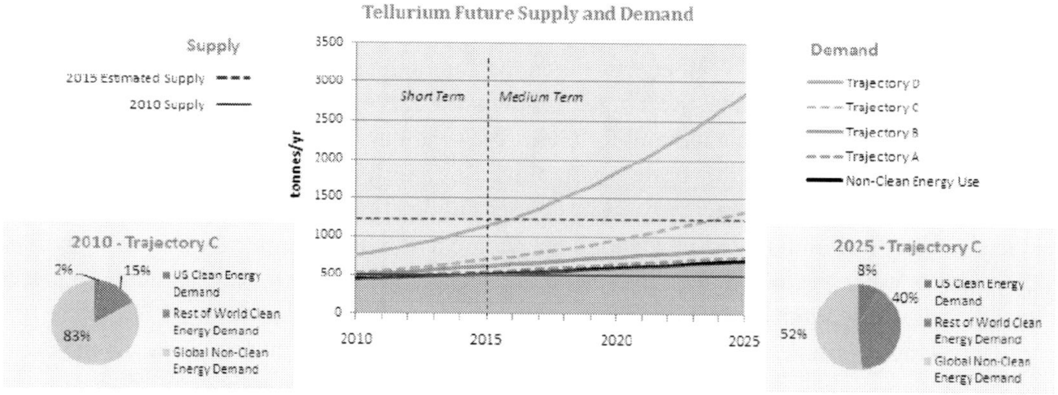

Figure 7-8. Future demand and supply for tellurium.

Figure 7-7 illustrates that the basic availability of cerium oxide appears to be more than adequate to meet demand for cerium in lighting phosphors and nickel metal hydride vehicle batteries. Demand for cerium oxide in nickel metal hydride batteries is more than 10 times higher than in lighting phosphors under Trajectory C in 2025. Non-clean energy dominates overall demand, accounting for 84% of global demand in 2025 under Trajectory C. Additional mines anticipated to come online by 2015 appear sufficient to meet demand until the middle of the medium term. To meet high-technology-growth trajectories (C and D), additional mines contributing a total of at least 10,000 tonnes per year will be needed. However, cerium oxide is commonly found in high concentrations in ore bodies and is unlikely to experience a supply-demand imbalance.

7.5. Trajectories of Future Demand for Key Materials in Thin-film Photovoltaic Power Systems

The use of key materials (e.g., indium, gallium and tellurium) in thin film PV power systems is described in Chapter 2. The thin film technologies considered are cadmium telluride (CdTe) and copper-indium gallium diselenide (CIGS). The assumptions for total PV deployment, market share of CdTe and CIGS modules and material intensity used to create trajectories of future demand for key materials in PV technologies are presented in Table 7-6.

The IEA WEO "**2009 Reference Case**" was selected as the basis for the *Low Penetration* case for photovoltaic power systems. Recall that this case assumes that no new policies will be implemented to accelerate adoption of these technologies after 2009. The IEA WEO "450 Scenario," meant to stabilize greenhouse gas concentrations at 450 ppmv, was used as the basis for the *High Penetration* case for PVs.

The development of the market share and material intensity assumptions is presented in detail in Appendix B.

Figure 7-8 illustrates the range of projections for tellurium demand over the period 2010–2025, considering both non-clean energy demands and demand for CdTe PV modules. Figure 7-9 and Figure 7-10 illustrate the ranges of demand for indium and gallium respectively through non-clean energy demand and for CIGS PV modules.

Figure 7-8 shows that the basic availability of tellurium appears more than adequate because expected increased recovery from copper anode slime dramatically increases supply in the short term. If anticipated new supplies become available, it appears that supply will be sufficient to meet projected demand past 2020 in Trajectories A–C. Reducing material intensity provides significant payoff for reducing overall material demand. If CdTe PV approaches the low material intensity of 43 tonnes per GW, only a minimal increase in supply in the medium term is necessary to accommodate high PV penetration. Non-clean energy demand as a percentage of total demand shrinks significantly over time under Trajectory C, but will still account for a large share of global demand in 2025. U.S. clean energy demand for tellurium is about a sixth of global clean energy demand in 2025, under Trajectory C. To meet global tellurium demand under Trajectory D, supply in 2025 must double estimated 2015 supply.

Figure 7-9 shows that the basic availability of indium appears somewhat tight by 2015, particularly for the trajectories with high material intensity. Without market adjustment, supply will need to increase by more than 25% over the 2015 estimate to meet just non-clean energy demand in 2025. Non-clean energy demand dominates indium consumption in Trajectories A–C. Clean energy demand adds to this demand, at 11% of total demand in 2025 under Trajectory C. Reducing the material intensity of indium in CIGS photovoltaic cells provides significant reductions of overall material demand. Without expanded production after 2015, reductions in non-clean energy demand will also be important to prevent shortages and price spikes.

Figure 7-10 shows that the basic availability of gallium appears more than adequate. Estimated 2015 supply is sufficient to meet projected demand beyond 2020 in all but Trajectory D. Continuing industry trends of reducing material intensity provides significant payoff for reducing overall material demand. While the share of clean energy demand relative to non-clean energy demand increases from 5% in 2010 to 16% in 2025, non-clean energy demand still dominates across Trajectories A–C. U.S. clean energy demand contributes about one-sixth of global clean energy demand in 2025 under Trajectory C. High penetration of CIGS PV without advances in material intensity, global supply of gallium would need to increase by roughly 85% between 2015 and 2025 to meet global demand.

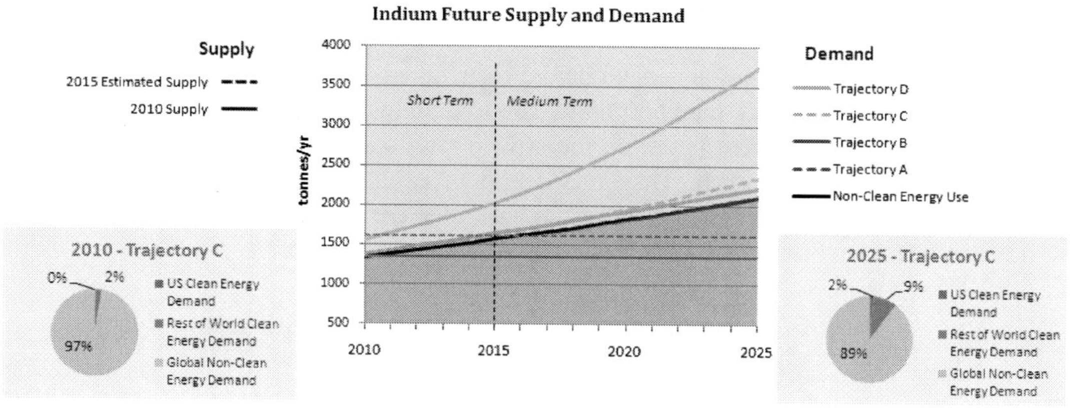

Figure 7-9. Future demand and supply for indium.

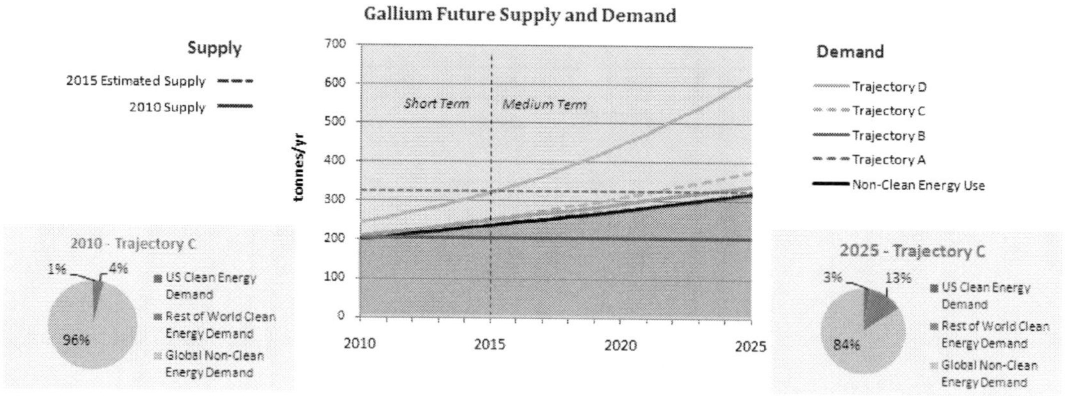

Figure 7-10. Future demand and supply for gallium.

7.6. Trajectories of Future Demand for Rare Earth Elements in Phosphors for High-Efficiency Lighting Systems

The use of different rare earth elements (including europium, terbium, gadolinium, cerium, lanthanum and yttrium) in phosphors is discussed in Chapter 2. High-efficiency fluorescent lighting represents approximately 85% of global demand for rare earth phosphors (although phosphors represent a small fraction of the total use of each rare earth element).[82] Phosphor demand will continue to grow with the increased use of high efficiency linear fluorescent lamps (LFLs) and compact fluorescent lamps (CFLs).

The assumptions for rare earth phosphors for lighting and material intensity used to create trajectories of future demand for key materials in lighting technologies are presented in Table 7-7.

Market share is accounted for in the deployment rate trajectories and not broken out as a separate assumption. No published global scenarios were identified that compared high and low penetration of high-efficiency lighting using phosphors based on REEs. As a consequence, the *High Penetration* and *Low Penetration* cases for rare earth phosphors in lighting applications were generated on the basis of assumed high and low compound annual percentage growth rates for phosphors. The low annual growth rate of 2.2% and the high growth rate of 3.5% were based on the growth rates for CFLs in the IEA 2010 "Phase Out Incandescent Lamps" study. Due to a dearth of credible data, there was only one set of assumptions used for material intensity of each element in this analysis. The development of material intensity assumptions is presented in detail in Appendix B.

Table 7-7. Assumptions for Key Materials in Lighting Technologies

	Technology	Assumption	Low Penetration	High Penetration
Deployment in 2025	Lighting	Rare Earth Phosphors for Lighting (tonnes)	**9307** (Based on 2.2% annual growth rate)	**11,250** (Based on 3.5% annual growth rate)
Market Share		Included in deployment assumptions		
	Technology	**Assumption**	**Low Intensity**	**High Intensity**
Materials Intensity	Lighting	% Weight of lighting phosphors that is Lanthanum	8.50%	8.50%
	Lighting	% Weight of lighting phosphors that is Cerium	11.00%	11.00%
	Lighting	% Weight of lighting phosphors that is Europium	4.90%	4.90%
	Lighting	% Weight of lighting phosphors that is Terbium	4.60%	4.60%
	Lighting	% Weight of lighting phosphors that is Yttrium	69.20%	69.20%

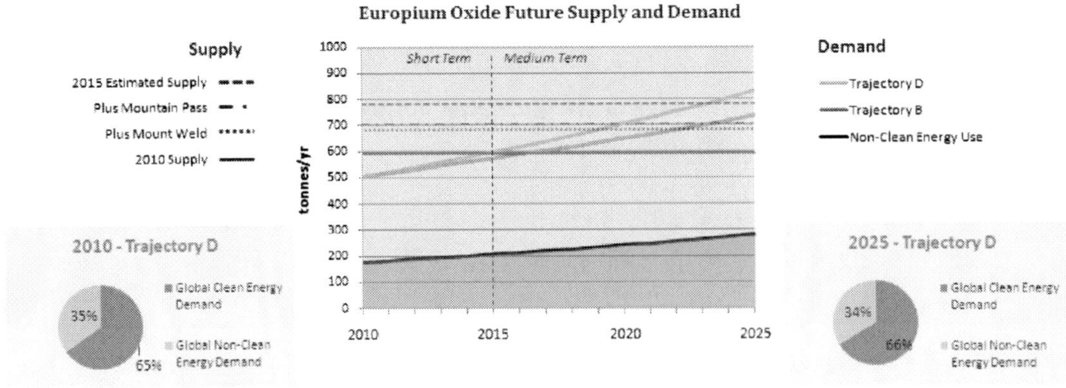

Figure 7-11. Future demand and supply for europium oxide.

Figures 7-11 to 7-13 illustrate estimated supply and the range of demands projected in this study for REEs used as phosphors in high-efficiency lighting technologies (except lanthanum and cerium, which were discussed earlier). These three figures contain only Trajectories B and D because only one set of material intensity values for phosphors was available. U.S. and global clean energy shares of total demand are based on Trajectory D (equivalent to Trajectory C for these materials).

Figure 7-11 shows that the basic availability of europium oxide is more than adequate in the short term and likely adequate in the medium term. Unlike most key materials, clean energy demand for europium oxide in phosphors dominates the overall demand and accounts for approximately two- thirds of global demand throughout the short and medium term. The addition of supplies from Mount Weld increases global supply by 15% and represents a significant portion of the increased production in 2015. Only demand under Trajectory D is projected to exceed 2015 supply by the end of the medium term. Trajectories assume limited substitution of light emitting diodes (LEDs) and organic light emitting diodes (OLEDs) for fluorescents, which could substantially mitigate demand in the 2020–2025 timeframe.

Figure 7-12 shows that the basic availability of terbium oxide is adequate in the short term, but may become tight early in the medium term without additional production after 2015. Both high- and low-growth trajectories will exceed forecast 2015 demand by the middle of the medium term. Mount Weld and Mountain Pass, which are projected to open in 2011 and 2012, respectively, have limited capacity to produce terbium oxide, so increased supply by 2015 will come largely from Nolans Bore mine in Australia. To meet demand under Trajectory D in 2025, supply will need to increase by 140 tonnes per year over 2015 estimated supply. Trajectories assume limited substitution of LEDs and OLEDs for fluorescents, which could substantially mitigate demand in the 2020–2025 timeframe.

Figure 7-13 shows that the basic availability of yttrium oxide appears very tight in the short and medium term. Global demand in 2010 exceeds current production.[84] Production is anticipated to grow 11% by 2015. At this level, it will be unable to meet growing demand in the short and medium terms. Significant additional production will be needed to meet short term demand. Production will need to grow by more than 40% over the 2015 level by 2025 in order to meet even the low-growth trajectory. Clean energy demand in phosphors accounts for approximately 50% of global demand throughout the short and medium term. Trajectories assume limited substitution of LEDs and OLEDs for fluorescents, which could substantially mitigate demand in the 2020–2025 timeframe. If high- temperature superconductors, which generally contain yttrium, begin to capture market share from permanent magnets for use in wind turbines, demand for yttrium may increase more rapidly.

7.7. Market Dynamics Affecting the Pricing and Availability of Rare Earth Elements and Other Key materials

Market dynamics refer to the interactions between demand, supply and prices. The market dynamics that affect REEs and other key materials vital to the commercialization of clean energy technologies are not captured by traditional economic models or simple economic analyses. This section discusses how factors associated with the extraction, processing and use of these materials can lead to supply-demand imbalances that manifest either in shortages or large price fluctuations. Also included is how the same factors make it difficult for supply and demand to respond to price signals.

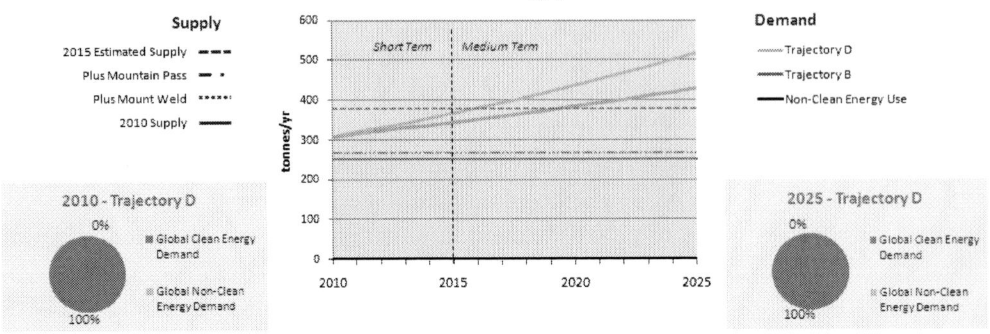

Figure 7-12. Future demand and supply for terbium oxide[83].

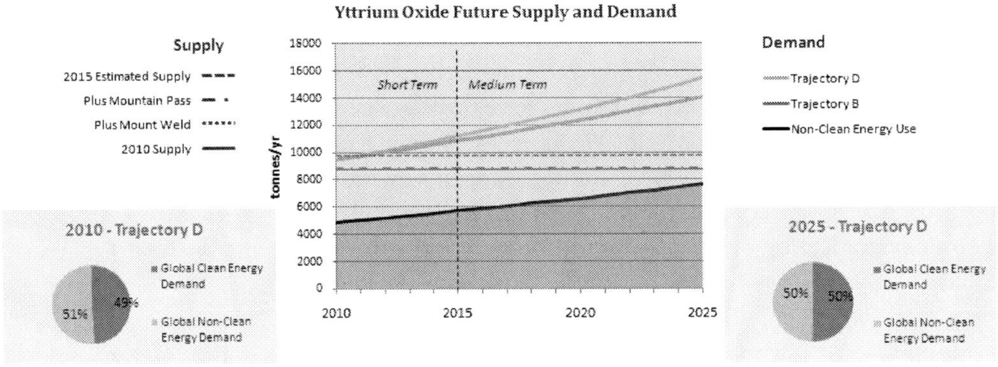

Figure 7-13. Future demand and supply for yttrium oxide.

Weak Price Signals

Rare earth elements and other key materials are not generally traded on spot markets. Instead, these materials are usually purchased via long-term, bilateral contracts between individual suppliers and manufacturers. These transactions provide irregular pieces of price information because they are large and cover long periods. The negotiated prices in bilateral contracts are influenced by many factors associated with the contracting parties, such as market power and credit rating, making it difficult to generalize price information. The price established in one bilateral contract is unlikely to be repeated in another contract due to these factors. This results in a lack of reliable price information for supply- and demand-side players to incorporate into business decisions. Weak price signals make it more difficult to take necessary actions to either increase supply or reduce demand.

Demand-side Factors

There are several demand-side factors that influence the market dynamics for REEs and other key materials. The most important demand-side factors include low value share, limited substitutes, challenge of reducing material intensity and policy uncertainty.

The key materials addressed in this report are used in relatively small amounts in components of clean energy technologies. The cost of a given REE or other key material input represents a small percentage of the total cost of a clean energy technology. Therefore, even large changes in the prices of these materials—as occurred for many REEs during the last decade and again during the summer of 2010— have a marginal impact on producer costs and the final consumer price.

REEs and other key materials are used in a wide array of technologies because of their extraordinary properties. For many technologies, there are limited substitutes available at any price that offer comparable performance characteristics. A few examples where there are limited substitutes include the use of cerium as a polishing compound, europium as a phosphor and neodymium in high-strength permanent magnets. Extensive research and development (R&D) may identify substitutes in the future, but individual firms on the demand side will be challenged to justify the expense. Only under high prices and certainty about sustained high prices will firms actively pursue R&D into substitutes.

Lowering the material intensity of REEs and other key materials in components for clean energy technologies also faces financial hurdles. Similar to pursuing substitutes, the business case to fund R&D into lowering material intensity in a clean energy technology depends on sustained and predictable high prices. Combined with low value share and limited substitutes, the expense of lower material intensity means that price increases have little effect on end-use demand. In economic terms, this is referred to as inelastic demand with respect to price.

Government policy is an important factor driving demand for clean energy technologies and the components containing REEs and other key materials. As seen in the figures earlier in this chapter, policy initiatives at the global and national level can drastically alter the deployment level of a given clean energy technology. Government support has the ability to help clean energy technologies overcome financial and other market hurdles. This complicates the market dynamics by creating uncertainty about future demand and by disconnecting demand from costs and available supply considerations.

Supply-side Factors

In addition to the demand-side issues, there are several supply-side factors that influence the market dynamics for REEs and other key materials. The main supply-side factors are coproduction, concentrated market power, large capital requirements and long lead times for mining projects.

Foremost among supply-side factors is the complexity of coproduction with other materials. None of the *individual* REEs or other key materials discussed above is mined individually or as the primary focus of commercial extraction activities. Instead, they are produced as coproducts or secondary products of other materials that may have entirely different applications, scale of use, prices and market demand. This creates the potential for supply of key materials to be completely or partially independent of demand or price levels for those key materials.

Table 7-8 below illustrates possible primary extraction products for some of the materials examined in the Strategy. In addition, even where rare earth elements are not byproducts of a primary extraction product, multiple rare earth elements are commonly coproduced with one another.

The impact of coproduction on market dynamics is a function of the prices of the coproducts, their abundance in the ore and the costs of separation and refining for each product. For example, the iron produced as a primary product in the Baotou (China) mines is less valuable by weight than the REE byproducts, but annual iron production represents more than 100 times the value of the bastenite ore. The iron is also more easily separated from the ore and processed into a marketable commodity. In cases like Baotou, mines may generate the majority of their revenue from extraction of the primary product, ignoring considerations of the minor metal and making operational decisions primarily based on the price and demand for the primary product.

Currently, China is the source of more than 95% of the world's rare earth oxides. This concentrates nearly all the supply-side market power in a single actor. This means that actions by China related to REEs, such as the recent tightening of export quotas, directly influence the market. For example, during the late summer and early fall of 2010, prices for many REEs increased 300%–700% (as shown in Chapter 3) in the wake of a series of geopolitical developments involving China. In addition, decisions concerning REE production in China may be motivated by factors beyond the REE market.

Table 7-8. Primary Extraction Products Related to Key Materials

Material	Primary Extraction Product
Rare Earth Elements	Iron
Cobalt	Nickel and Copper
Gallium	Aluminum and Zinc
Indium	Zinc
Tellurium	Copper

In addition to coproduction issues, producers of REEs and other key materials are limited in their ability to increase supply. Even if the market demand for a particular element would make increased production highly profitable on a tonnage basis, mining companies may be unable to rapidly increase production or quickly open new mines. They are constrained by capital costs typically in the billions of U.S. dollars and by long lead times required for exploration, permitting and facility construction. Some industry experts estimate that a 10-year lead time from initial exploration to the construction of new mines for rare earth elements is typical of the industry. Many of the enterprises working to develop rare earth reserves are junior mining companies engaged in only one project and do not have the financial resources to bring their ore bodies into production. As a consequence, for the developer to achieve commercial viability the company in many cases must "lock in" a buyer for the output of the mine even before the separation and refining process is fully operational. This creates enormous management challenges for the developers and substantial uncertainty for both the investors in the mine and the potential buyers of its output. Volatility in mineral prices also play a key role in these decisions; mining companies and their investors may be reluctant to invest in new mines without a reasonable expectation that prices will remain high enough over the full design life of the proposed mine to secure an attractive rate of return on their initial capital investment.

In conclusion, market dynamics for REEs and other key materials are complicated. These markets are far from perfectly functioning competitive markets from economic theory. Price signals for key materials tend to be muted by the lack of price transparency. Even if price signals were clear, the ability of supply and demand to respond to price changes is quite limited. The result is that shortages and significant price fluctuations are likely in the future, though very difficult to predict. Due to the complexities, no attempt was made in this study to forecast key material prices.

7.8. Conclusion

Four clean energy technologies have been the principal focus of this analysis:

- Permanent magnets made from alloys of REEs used in wind turbines and advanced vehicles with electric drive-trains
- Advanced batteries that incorporate REEs in their electrodes or are based on lithium-ion chemistries used in advanced vehicles with electric drive-trains
- Photovoltaic power systems using thin-film semiconductors
- Rare earth phosphors used in high-efficiency fluorescent lighting systems

Efforts to accelerate the commercialization and deployment of these four clean energy technologies face considerable risks of supply-demand imbalances that could lead to increased price volatility and supply chain disruption. The character and severity of these risks varies among the REEs and other key materials evaluated in this study. A number of options are available to the Department of Energy to reduce or mitigate these risks. These options will be discussed and evaluated in the following chapters of this study.

8. Criticality Assessment

Short- and medium-term criticality assessments of the various key materials (identified in Chapter 1) address two dimensions: importance to clean energy and supply risk. The basic premise is that rapidly increasing demand for key materials could hamper the clean energy agenda by outpacing new mining projects and causing supply/demand mismatches. Detailed element-by-element assessments are presented in Appendix A.

8.1. Assessment Methodology

The National Academy of Sciences (NAS) "Minerals, Critical Minerals, and the U.S. Economy" study (NAS 2008) developed a conceptual methodology to assess the criticality of individual minerals along two dimensions: "Impact of Supply Disruption" and "Supply Risk." These two dimensions are rated on a scale from one to four and presented on a matrix to visually communicate the relative criticality of individual minerals. According to this scheme, the upper right-hand corner of the matrix attains highest criticality.

The NAS methodology has been adapted in several ways for this Strategy. First, because the purpose of this Strategy is to address materials in clean energy, the "Impact of Supply Disruption" has been reoriented to become "Importance to Clean Energy." Second, there have been some adjustments to the attributes used to characterize "Supply Risk." Third, assessments have been completed for both short- and medium-term criticality, as these two time horizons have different supply and demand profiles and also different policy options.

Analogous to the NAS methodology, the two-dimensional criticality ratings are plotted on a matrix to enable comparison across materials for both short and medium term. The matrices inform a comparison among materials that can feed into prioritized research and development (R&D) investment and policy action. Each matrix has three regions: critical (red), near-critical (yellow) and non-critical (green).

Short- and medium-term scores for "Impact of Supply Disruption" and "Supply Risk" are based on a weighted average of two attributes. For each attribute, key materials were assigned qualitative factor scores of 1 (least critical) to 4 (most critical). The attributes are described in more detail below.

Importance to Clean Energy

Importance to clean energy encompasses two attributes for each material over the short and medium term. The weighting factor for each attribute is shown in parentheses.

Clean Energy Demand (75%): captures the importance of the material in magnets, batteries, photovoltaic (PV) films and phosphors used in clean energy technologies.

Substitutability Limitations (25%): addresses constraints on practically substituting for the material and technology within clean energy technologies. Substitution could occur at any level of the supply chain. This may include using different raw materials, components or even end-use technologies. This includes substitution by element, such as mischmetal for lanthanum in batteries, and also component technology-based substitutions, such as induction motors for permanent magnet motors.

Supply Risk

The overall supply risk for each material is based on five categories of risk for the short and medium term. For each category, key materials were assigned qualitative factor scores of 1 (least critical) to 4 (most critical). The categories are described in more detail below.

Basic Availability (40%): the extent to which global supply will be able to meet demand. Short-term basic availability examines mine and other production relative to demand. Medium-term basic availability examines the potential for other mines to begin producing the material relative to anticipated increases in demand. The qualitative score is informed by the projections in Chapter 7, but may also take into account other factors such as global reserves, mines projected to start up after 2015 and additional supplies from recycling.

Competing Technology Demand (10%): captures whether non-energy sector demand is expected to grow rapidly, thus constraining the supply of the material available for the energy sector.

Political, Regulatory and Social Factors (20%): risk associated with political, social and regulatory factors within major producer countries. This includes the risk that political instability in a country will threaten mining and processing projects; that countries will impose export quotas or other restrictions; or that social pressures, permitting and regulatory processes will delay the start up of new mines.

Co-dependence on other Markets (10%): covers instances where a mineral is coproduct or byproduct of with other minerals found in the same ore deposit. Co-dependence can be an advantage or a disadvantage, depending on which mineral is driving production levels overall. In general, coproducts with lower revenue streams (i.e. production rate X price) will have higher scores since they are less likely to drive production than coproducts with higher revenue

Producer Diversity (20%): captures market risks due to the lack of diversity in producing countries or companies (e.g., monopoly or oligopoly).

8.2. Identification of Critical Materials

Assessment scores for each key material were developed using the best available information and are shown in detail in Appendix A. The overall scores for importance to clean energy and supply risk are plotted in Figures 8-1 (short term) and 8-2 (medium term). Figure 8-3 shows the movement of scores from the short to medium term. Each of these plots can be thought of as a criticality matrix. Note that, in general, the criticality of some materials changes over time, due in some cases to anticipated market response and the emergence of viable substitutes on the one hand or a dramatic ramp up in demand for the material on the other. It is important to keep in mind that these are qualitative assessments, informed by some quantitative analyses. There is much uncertainty in the attributes examined, particularly in the medium term. While the collection of assessments is valuable to inform policy action and R&D investment, it will be important to revisit the analyses moving forward as more data are available and as material supply and demand changes.

The criticality matrices in Figures 8-1 and 8-2 suggest three broad categories of criticality. Materials in the upper quadrant of the chart (with scores of three or higher on both axes) are characterized as critical. Materials with a score of three or higher on one axis but a two on the other axis are characterized as near-critical. While they are not currently judged to be critical, small changes in one or more risk category could put them at criticality. All other materials are judged not to be critical. However, all of the assessments are based on the best available information, so even materials judged not critical could be at risk due to significant unforeseen circumstances. The distribution of materials by criticality categories in the short and medium term is shown in Table 8-1.

According to the analysis, indium and rare earth elements dysprosium, terbium, europium, neodymium and yttrium are critical in the short term. The uses for the critical rare earth elements are spread across magnets, batteries and phosphors and indium is used in PV films. Thus, each of the clean energy technologies examined in the Strategy has at least one critical material in the short term. Tellurium, cerium and lanthanum are near-critical, and gallium, lithium, cobalt, praseodymium and samarium are not critical. Between the short and the medium term, importance to clean energy and supply risk shift for some materials. For example, the supply risk for neodymium decreases, while the importance to clean energy increases. The supply risk for dysprosium and yttrium both decrease. On the other hand, both the importance to clean energy and the supply risk for lithium increase to make lithium near critical in the medium term.

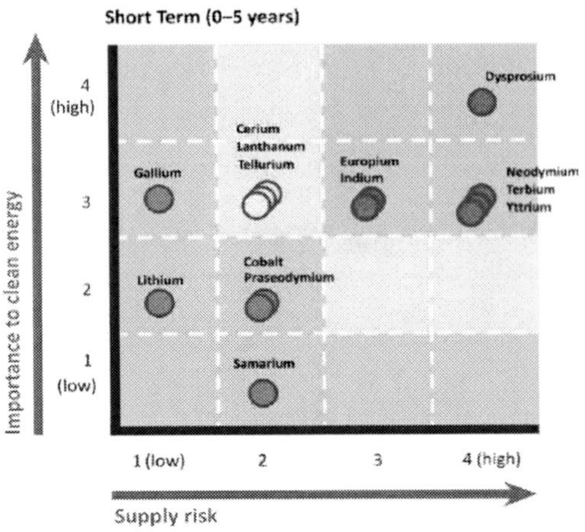

Figure 8-1. Short-term (0–5 years) criticality matrix.

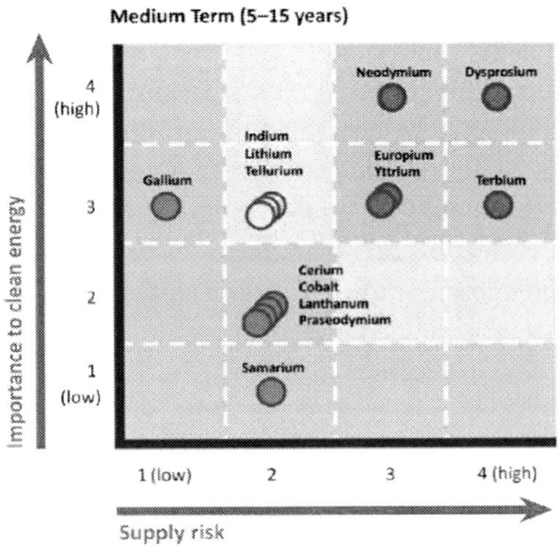

Figure 8-2. Medium-term (5–15 years) criticality matrix.

Table 8-1. Distribution of Materials by Criticality Category

Short Term	Medium Term
Critical	**Critical**
Dysprosium Europium Indium Terbium Neodymium Yttrium	Dysprosium Europium Terbium Neodymium Yttrium
Near-Critical	**Near-Critical**
Cerium Lanthanum Tellurium	Indium Lithium Tellurium
Not Critical	**Not Critical**
Cobalt Gallium Lithium Praseodymium Samarium	Cerium Cobalt Gallium Lanthanum Praseodymium Samarium

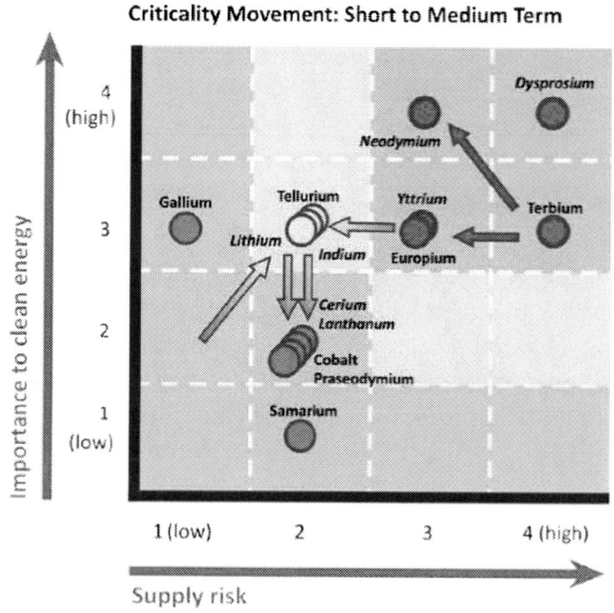

Figure 8-3. Comparison of short- and medium-term criticality.

Lithium is the only key material that shifts into a higher criticality category from the short to medium term. This change is due to the rapid increases in market penetration projected for vehicles using lithium-ion batteries, which increases lithium's importance to clean energy. This market penetration would significantly increase demand even as lithium production capacity increases, thus increasing supply risk slightly.

All other key materials either remain in the same category or become less critical in the short to medium term. For the materials that shift to a lower category, this change generally reflects a combination of expanded supply and increased alternatives for substitution at different levels of the supply chain. Market dynamics, described in Chapter 7, will play a large role in these positive criticality changes. However, that same discussion also highlighted the market complexities and distortions that remain for many key materials, suggesting an important role for government. The next chapter lays out the U.S. Department of Energy's strategy for using targeted research and development, combined with other policy options, to address material criticality.

9. Program and Policy Directions

In preparing this Strategy, the U.S. Department of Energy (DOE) considered programs and policies in eight broad categories: (i) research and development, (ii) data collection, (iii) permitting for domestic production, (iv) financial assistance for domestic production and processing, (v) stockpiling, (vi) recycling, (vii) education and (viii) diplomacy. These address risks, constraints and opportunities across the supply chain as shown in Figure 9-1. DOE's authorities and historic capabilities with respect to these categories vary widely. Some (such as research and development) relate to core competencies of DOE. Others (such as permitting for domestic production) concern topics on which DOE has little or no jurisdiction. These programs and policies address risks, constraints and opportunities across the supply chain, as shown in Figure 9-1. A discussion of each of these topics follows.

9.1. Research and Development

As the nation's leading funder of research on the physical sciences, DOE's capabilities with respect to materials research are substantial. The DOE national laboratory system includes the nation's historic leader in rare earth metal research—the Ames National Laboratory in Ames, Iowa. Other DOE national labs also conduct materials research. DOE programs supporting materials research include the Office of Energy Efficiency and Renewable Energy (EERE), Office of Science and Advanced Research Projects Agency-Energy (ARPA-E).

During November and December 2010, DOE convened three technical workshops on rare earth metals and other critical materials. A U.S.-Japan workshop was held November 18th and 19th at Lawrence Livermore National Laboratory; a U.S.-European Union workshop was held December 3rd at Massachusetts Institute of Technology; and ARPA-E held a workshop in Washington, D.C., December 6th.

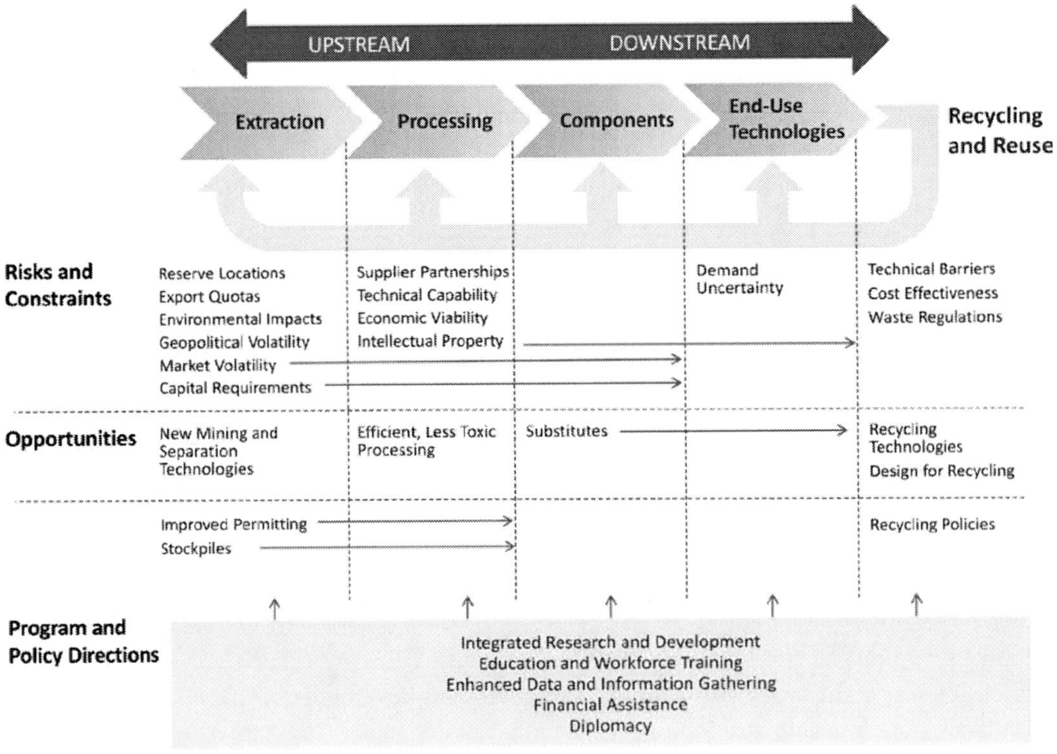

Figure 9-1. Policy options and the critical material supply chain.

In early 2011, DOE will evaluate the results of these workshops and develop the Department's first integrated research plan with respect to critical materials. This research will build on existing work at DOE and elsewhere. In addition to addressing potential synergies across DOE, the research plan will also consider opportunities to collaborate with other agencies and departments, including the Department of Defense (DoD), National Science Foundation, Department of the Interior and Environmental Protection Agency.

Objectives for DOE's research will include (i) reducing the materials intensity of clean energy technologies and (ii) cutting costs and improving environmental performance across the full supply chain. Sustained and integrated R&D may provide breakthroughs in substitute materials – or substitute technologies—that significantly reduce materials intensity, thus decreasing import dependence and exposure to supply interruption. Additionally, R&D into recycling, design for recycling and more efficient use will help maximize available supply. Accomplishing these goals may involve R&D in fundamental materials development, manufacturing process improvements, systems design and integration advances and recycling process improvements.

Based on input to date, the Department expects that its integrated plan will give priority to the following topics:

Magnets, Motors and Generators

Rare earth permanent magnets play a vital role in a number of clean-energy technologies that convert electricity to and from kinetic energy, such as vehicle motors and wind turbines. However, there are opportunities to reduce the amount of rare earth metals used in magnetic systems while maintaining equivalent performance. There may also be opportunities to move away from the use of permanent magnets, although this will require either additional innovation or design trade-offs in weight and other parameters.

High-priority research areas include the following:

- Materials
- Nano-structured permanent magnets, including core-shell structures and composites
- Improved high-temperature performance of NeFeB magnets
- Enhanced magnetic coercivity for rare-earth, alnico and other magnets
- Fundamentals of anisotropy and new anisotropic mechanisms
- High-flux soft magnets;
- Molecular design of magnets
- Manufacturing
- Adapting advanced casting methods to enhance magnetic performance of alloys
- Improved process control to minimize waste
- Systems
- Optimized thermal management to reduce need for high-temperature-tolerance
- Optimized motor and turbine geometries to reduce friction and other operational losses

The Office of Science, EERE and ARPA-E are currently investing in magnet R&D. In addition, EERE and ARPA-E have been exploring collaboration with the Defense Advanced Research Projects Agency (DARPA) on this topic.

Batteries, Photovoltaics and Lighting

DOE currently supports robust research programs in batteries, photovoltaics (PV) and lighting— three important areas of the clean energy economy. Materials considerations will continue to be integrated into these research programs. The high-priority research areas include the following:

- Batteries
- Continued R&D supporting advanced technologies that utilize abundant elements, such as iron and zinc
- Photovoltaics
- Improved deposition processes to reduce cadmium telluride (CdTe) and/or copper-indium gallium diselenide (CIGS) active layer thickness and minimize deposition waste
- Materials research on polycrystalline PV alternatives
- Lighting
- Alternative phosphor materials, including the use of quantum dots that minimize or eliminate the use of cadmium or rare earth elements—the most difficult materials issues in lighting are likely to revolve around the supply of terbium (used for green phosphors) and europium (used for red and blue phosphors)
- Organic LEDs, with improvements to luminous efficacy, cost and color rendering

Environmentally Sound Mining

Research on low cost, environmentally sound mining, in collaboration with other agencies such as the Department of Interior and Environmental Protection Agency, could facilitate cleaner production at home and abroad. High-priority research areas include the following:

Non-traditional water source use (e.g., treated waste water, saline aquifer)

Long-term interactions between groundwater and mine excavations.

Alternatives to tailing impoundments

Optimized blasting and efficient crushing for lower energy consumption

Materials Processing

Innovations in the processing of ore into metals or other compounds are important for both the rare earth element and lithium supply chains. Processes with improved oxide or carbonate recovery, reduced energy consumption, reduced use of toxic agents and recycled or fully captured waste streams are most desired.

For rare earths, research could take advantage of such innovations as molecular design of solvent extraction reagents, advanced ion exchange, high-performance organic modifiers and advanced liquid membranes leveraging substantial U.S. knowledge base for heavy element chemistry. This heavy element chemistry scientific expertise has been developed for actinide chemistry in support of environmental management programs of DOE. Improved processes for extracting rare earths from mining tailings would also be beneficial. Valuable process enhancements for lithium include the energy-optimized conversion of ore to water-soluble lithium sulfates.

Recycling Research and Development

The materials examined in this Strategy are often not recycled, in part because they are used in small quantities in many technologies, both on a total and a per-unit basis. Additionally, per kilogram market prices are generally low relative to precious metals, so recycling is often not cost effective. However, as the use of these materials increases in vehicles and other common technologies, recycling could make more economic sense. Recycled content could become valuable as a secondary source on the market, which can ease periods of tight supply. Relevant research includes the following:

- Technology, component and material design for disassembly and recycling
- Collection, logistics and reverse supply chain optimization
- Recycling process development
- Recycling and reconditioning rare-earth materials from spent fluorescent lamps (with particular attention to safe and economical disposal of mercury)
- Recycling and reconditioning rare-earth materials from manufacturing yield loss
- Methods for efficient demagnetization of rotating-machine components
- Metallic flux processes for recovering rare earths

9.2. Data Collection

In developing this Strategy, DOE encountered a number of data gaps. Limited information is available, for example, with respect to the following:

- Annual production and consumption of individual rare earth metals
- Prices at which some rare earth metals trade
- Materials intensity of different energy technologies
- Potential for substitutes in different energy technologies, where critical materials are used

Good data contributes to sound policy- and decision-making. Reliable information is the basis for understanding the market situation, crafting goals and devising strategies to meet those goals. Data gaps make it challenging to assess material criticality, characterize markets and assess available technologies. Gaps in publicly available information also reduce the market's ability to self-correct while complicating private sector planning. Finally, the data gaps will limit the future ability of DOE and its interagency partners to assess the effectiveness of policies pursued today.

For material supply data, the U.S. Geological Survey (USGS) is the U.S. government's leading expert. Indeed USGS publicly available data on material supply, demand and pricing were invaluable in preparing this Strategy. Future analyses of critical materials will also depend on USGS data and information.

The Energy Information Administration (EIA), an independent agency within DOE, is the nation's premier source of publicly-available energy information. EIA collects, analyzes and disseminates independent and impartial energy information. Enhancements to EIA's existing data collection could significantly contribute to understanding of market dynamics with respect to critical materials. With appropriate resources, survey forms could be modified to collect key material content of various clean energy technologies.

Other parts of the Department of Energy, including the Office of Energy Efficiency and Renewable Energy and Office of Policy and International Affairs, also have expertise and maintain relationships that could be helpful in gathering relevant information to inform DOE planning.

In the months ahead, DOE will engage a broad range of stakeholders to better understand current trends, constraints and opportunities with respect to critical materials. DOE may hold additional technical workshops and/or issue additional Requests for Information related to these topics. DOE is considering developing periodic demand scenarios across clean energy technologies for individual rare earth elements as well as for lithium, cobalt, gallium, tellurium, cobalt and perhaps other key materials. The objective of these activities would include creating a more comprehensive understanding of the global market, providing more certainty to the private sector and facilitating investments. In all cases, DOE's plans with respect to additional data collection are subject to the availability of appropriated funds.

9.3. Permitting for Domestic Production

The Department of Energy does not regulate mining or mineral production. As a result, issues with respect to expedited permitting for domestic production of critical materials are mostly outside the scope of this Strategy.

However, the Department has a strong interest in ensuring diverse sources of supply for the critical materials necessary for clean energy technologies. Production within the United States is vitally important in that regard, for at least two reasons. First, the United States' considerable reserves of some critical materials could add significantly to total global production and to greater diversity in the global supply of these materials. Second, U.S. technology and best practices developed during mine operations can help promote safe and responsible mining in other countries, further contributing to supply diversity and the sustainable development of resources.

DOE will work with interagency colleagues where possible and appropriate to improve permitting of mines for critical materials in the United States while ensuring that worker safety, environmental protection and other important values are fully protected. In that regard, the Department notes obtaining the permits necessary to open a mine takes on average 7–10 years in the United States. Federal, state and local agencies all play a role. [85] These permits serve important public purposes, including protecting employee health and safety, protecting air and water quality, ensuring proper handling and disposal of radioactive substances and providing for remediation of soil and groundwater contamination. However this 7 to 10 year period is reportedly the longest among the top 25 mining countries (Behre Dolbear 2010). In Australia, by contrast, approvals take on average 1 to 2 years (Matthews 2010).

Options to simplify permitting may include improved coordination between state and federal agencies as well as among federal agencies during all stages of permitting.[86] In addition, government engagement with the private sector on best practices will better educate miners and prospectors before the permitting process begins. This can further accelerate mine development. Since the permitting process is often cited as one of the principal barriers to new mining ventures in the United States, a more coordinated and predictable regulatory process could encourage investment in new mines and eventually contribute to diversifying the global supply chain.

9.4. Financial Assistance for Domestic Production and Processing

DOE lacks authority to provide financial support for the domestic production or processing of critical materials. Such authority is not provided under DOE's Loan Guarantee program (Title 17 of the Energy Policy Act of 2005, as amended) or Advanced Technology Vehicle Manufacturing Program (section 136 of the Energy Independence and Security Act of 2007), nor is such authority provided under tax credit programs administered by Secretary of Treasury in consultation with DOE (including the Advanced Energy Manufacturing Tax Credit under Internal Revenue Code section 48c).

Loan Guarantee Program

The Department of Energy is authorized to issue loan guarantees for several purposes. Under current law, DOE may issue loan guarantees for new or significantly improved technologies that reduce air pollutants including greenhouse gases (Section 1703 of Energy Policy Act of 2005); for renewable energy systems, including the manufacture of renewable energy system components (American Recovery and Reinvestment Act of 2009, amending Section 1705 of the Energy Policy Act of 2005); and for the manufacture of advanced technology vehicles in the United States (Section 136 of the Energy Independence and Security Act of 2007).[87]

Each of these authorities may be available to support the use of rare earth metals or other critical materials for the primary purpose outlined in the statute. For example, DOE could issue loan guarantees for the domestic manufacture of magnets using rare earth metals, if those magnets will be part of renewable energy systems. DOE also has authority to issue loan guarantees for new or significantly improved technologies that reduce air pollution including greenhouse gases. This could apply, for example, to the processing of mineral ores including rare earth metals. However, DOE lacks authority to provide loan guarantees more generally for the mining or processing of rare earth metals or other critical materials.

The United States has a strong national interest in enhancing capacity for domestic production and maunfacturing of critical materials, which will help diversify the supply chain for certain materials.

Price Supports

One barrier to establishing and sustaining domestic production capacity for critical materials may be the ability of countries with considerable market share to cut prices, potentially driving mines in other countries out of business. If there is a national interest in the United States developing the capacity to produce rare earth metals and other critical materials domestically, it may be worth analyzing whether some type of price support system is appropriate. More data collection and analysis would be required to define the parameters of such a program, determine its potential costs and consider conditions under which it might be in the national interest.

9.5. Stockpiles

Several countries have developed or are considering stockpiles for key materials, as described in Chapter 6. In addition, the DoD has moved to include many of the materials considered in this Strategy in a new Strategic Military Stockpile Program (SMSP), as described in Chapter 5. The SMSP will help meet the DoD's legal requirement to provide special monitoring and attention for all strategic materials and ensure that those materials further deemed as critical to national security are also sourced domestically (Defense Strategic Materials Protection Board 2008).[88]

In theory, stockpiles could protect the United States from interruptions in supply from foreign producers. The existence of a stockpile, whether or not used, could have geopolitical significance, diminishing the leverage of monopoly suppliers in crisis situations. If coupled with purchase and price guarantees, stockpiles might also promote investment in new domestic mines by providing protection from commodity price swings.

However, stockpile authorities must purchase material that would otherwise be sold on the open market. This could be expensive and itself have market impacts, increasing the price of purchased commodities. Additions to stockpiles could induce shortages or price spikes if not done carefully.

Based on preliminary analysis, this Strategy does not recommend stockpiling critical materials for potential use in commercial clean energy technologies at this time. The demand projections for material use in clean energy technologies presented earlier in the Strategy highlight the difficulties in accurately forecasting material requirements due to uncertainties in market conditions, choice of component technologies among manufacturers and competing demands. From a practical standpoint, these factors would make it difficult to develop a national industrial stockpile with sufficient material stocks and flexibility. Even if material requirements could be calculated with a reasonable degree of certainty, the U.S. Government would incur significant upfront costs and downside risk to develop a stockpile sufficient to meet domestic material demand. Maintaining a national stockpile would also put the government at risk of distorting market price signals for key materials by competing with the private sector for materials on the open market. However, given the demonstrated interest of other nations, such as China, in stockpiling, this issue merits further study.

In addition, although stockpiling materials for clean energy technologies is not currently recommended, the DOE should work in close coordination with the DoD to understand and inform the development of the SMSP. This coordination would help identify areas where strategic materials requirements for military and other national security applications overlap with material requirements for clean energy technologies, to ensure that any actions by the SMSP do not adversely impact the development and deployment of clean energy technologies.

In the short term, DOE can encourage industry to increase private stocks and inventories, to the extent practicable and possible, in order to maintain resiliency in case of future supply disruptions. Private sector stocks of critical materials can then be traded between market participants to help balance supply and demand. Spreading stocks among participants will help buffer the market and provide necessary market adjustments. Such an approach would have to be applied carefully in order to avoid inducing or exacerbating shortages.

9.6. Recycling Policy

Recycling can reduce the risk that supplies of critical materials fail to keep pace with demand. Recycling may be especially important for critical materials with limited substitutes. If done correctly, recycling can also reduce environmental impacts from disposal of end-of-life products and equipment (LaMonica 2010). Historically, few companies have recycled the materials examined in this Strategy because there was little economic incentive to do so. First, recoverable quantities were often too small and raw material prices too low, for recycling to be economically attractive. For example, there is currently little economic incentive for businesses to recover lithium from lithium- ion batteries in consumer electronics because the quantities found in each battery are small and the price of lithium is low compared to most other metals (Hamilton 2009). However, the economic equation may change for the larger lithium-ion batteries likely to be used in electric vehicles. Second, expensive and/or energy-intensive processes have often been required to recycle and separate materials to desired purity levels (Ames Laboratory 2010). Third, in many cases, there is not an efficient mechanism for the collection of end-of-life products and equipment (Umicore 2010).

Policies directed toward the recovery of end-of-life products and equipment can encourage a higher rate of recovery. Supporting research and development into more efficient and cost-effective recycling processes can also make recycling more attractive. In addition, government assistance can support the development of recycling infrastructure, particularly where there may be a growing need, such as for electric vehicle batteries. DOE will work with interagency colleagues to analyze these and other options.

9.7. Education and Workforce Training

As the domestic industry in rare earth metals and other critical materials grows, a trained workforce will be increasingly important. Today, employment opportunities in these areas are limited, in part because of the small size of the sector. Yet the sector is less likely to grow without trained workers. Investment in education and training, alongside investment in productive capacity, can help support the country's manufacturing base. A robust education and training system can help spur innovation. Cooperation among government, industry and research institutions can be important drivers of clean energy innovation.

In the years ahead, materials sciences will receive increasing attention in DOE's internships, fellowships and scholarships. It is hoped that universities and corporate research labs will also encourage students to engage in research on rare earth elements and other key materials—. Postdoctoral fellowships can help train the next generation of materials scientists and engineers in these important areas. Expertise that would be valuable includes mineral and mining engineering, mineral economics, materials recycling technology and manufacturing engineering (NRC 2008). Finally, DOE will work with interagency partners such as DoD and the National Science Foundation to highlight the need for expertise in material sciences and help create opportunities in the public and private sectors. The strength of the U.S. industry in this sector depends on sufficient human capital to support mining and processing operations, as well as promote the innovations needed across the supply chain.

9.8. Diplomacy

Many other countries are considering questions similar to those explored in this Strategy. Cooperation with those countries can provide useful information and help improve transparency in markets for critical materials. It can also help to optimize resources for research and accelerate research and development on key topics. For these reasons and more, DOE will work closely with counterparts in other governments in the years ahead.

More broadly, addressing these issues will require sustained U.S. diplomatic engagement. Working closely with the State Department and other agencies, DOE will engage other countries through dialogues and collaborative institutions. A primary objective will be to maintain frequent and open communication with important stakeholders. Building on ongoing discussions with partners such as the European Union and Japan, DOE will engage significant producers and consumers of critical materials. DOE will participate in multilateral fora, such as the International Energy Agency, to advance our goals with respect to critical materials as well. On issues of trade promotion and compliance, DOE will support the U.S. Trade Representative in its efforts to uphold the rules-based global trading system and ensure open and fair global markets for producers and consumers of critical materials. As other nations pursue aggressive strategies to secure valuable resources, DOE can encourage others to avoid market manipulation and ensure a level playing field for all users.

As a dominant market player in a number of critical materials, China will be an important interlocutor. It is in the interest of both China and the United States to promote globally diverse, sustainable and economical supplies of clean energy materials for future use by both countries. Building on its existing energy diplomacy, DOE intends to work with China and other countries on these issues in the years ahead.

9. 9. Conclusions and Next Steps

As the analysis in this report demonstrates, widespread global deployment of clean energy technologies will likely change the future pattern of material consumption in those technologies. This change may include a substantial increase in the demand for some critical materials with limited basic availability and limited diversity of supply over the short and medium term. Left unaddressed, this reality will severely hamper the United States' ability to transition to a clean energy economy.

This Strategy presents three main goals for the critical materials market to begin addressing these challenges:

- Achieve globally diverse supplies
- Identify appropriate substitutes
- Improve recycling, reuse and more efficient use of critical materials

Achieving these goals will help mitigate supply risks and place the U.S. clean energy economy on a reliable and sustainable pathway. As a next step, DOE will work with U.S. Government partners to develop integrated programs and policies to achieve these goals. Possible policies include financial assistance for domestic processing and manufacturing, simplified permitting for domestic production, recycling and diplomacy.

Building on its deep technical expertise and the results of three workshops held on these topics in November and December 2010, DOE will develop its first integrated research plan with respect to critical materials. Priority attention will be devoted to magnets, batteries, PV films and phosphors. Innovations in environmentally sound extraction and materials processing will also be explored. DOE will also contribute to market transparency through enhanced data collection.

Investments in research, education and training with respect to critical materials can help strengthen the U.S. economy, creating jobs in many sectors. Innovations in magnets, recycling and material processing, will likely have applications beyond the energy sector. Environmentally sound mineral extraction and materials processing can contribute to local acceptance of these activities, as well as reduce costs over the long term.

Taken collectively, these programs and policies represent a robust first response to the challenges identified in this report. They provide meaningful benefits at each stage of the critical materials supply chain and help advance U.S. efforts toward diverse supply, substitution and improved recycling and efficient use. Of course, these efforts are just the beginning of a long-term approach to an important issue that will require sustained attention and coordinated action in the years ahead.

APPENDIX A: CRITICALITY ASSESSMENTS BY ELEMENT

This appendix provides the detailed assessments of criticality for each of the key materials. The methodology used to develop the criticality scores was explained in Section 8.1. For each material, the scores for "importance to clean energy" and "supply risk" are based on weighted averages of a number of individual factor scores. The descriptions of each factor were also presented in Section 8.1. Table A-1 summarizes the assessment scores for each key material in both the short and medium terms.

Table A-1. Short and Medium Term Criticality Scores for Key Materials

		Weight:	0.75	0.25	Weight:	0.4	0.1	0.2	0.1	0.2
	Atomic #	**Importance to Clean Energy (Rounded Score)**	**Clean Energy Demand**	**Substitutabilty Limitations**	**Supply Risk (Rounded Score)**	**Basic Availability**	**Competing Technology Demand**	**Political, Regulatory, and Social Factors**	**Co-dependence with other markets**	**Producer Diversity**
Short Term										
lithium	3	**2**	2	2	**1**	1	1	1	1	1
cobalt	27	**2**	2	2	**2**	1	2	3	2	2
gallium	31	**3**	3	2	**1**	1	2	1	3	1
yttrium	39	**3**	3	4	**4**	4	2	4	2	4
indium	49	**3**	3	2	**3**	4	3	1	3	1
tellurium	52	**3**	3	2	**2**	3	2	1	3	1
lanthanum	57	**3**	3	3	**2**	2	2	3	2	3
cerium	58	**3**	3	2	**2**	1	2	3	2	3
praseodymium	59	**2**	2	1	**2**	2	1	3	3	3
neodymium	60	**3**	3	3	**4**	4	2	4	2	4
samarium	62	**1**	1	1	**2**	2	1	3	3	3
europium	63	**3**	3	4	**3**	3	2	4	3	4
terbium	65	**3**	3	3	**4**	4	2	4	4	4
dysprosium	66	**4**	4	3	**4**	4	2	4	3	4
Medium Term										
lithium	3	**3**	3	2	**2**	2	2	1	1	1
cobalt	27	**2**	2	2	**2**	1	2	2	2	2
gallium	31	**3**	3	2	**1**	1	3	1	3	1
yttrium	39	**3**	3	4	**3**	3	2	3	2	3
indium	49	**3**	3	2	**2**	3	4	1	3	1
tellurium	52	**3**	3	2	**2**	3	2	1	3	1
lanthanum	57	**2**	2	3	**2**	2	2	3	2	2
cerium	58	**2**	2	2	**2**	1	3	3	2	2
praseodymium	59	**2**	2	1	**2**	2	2	3	3	2
neodymium	60	**4**	4	3	**3**	3	2	4	2	3
samarium	62	**1**	1	1	**2**	2	2	3	3	2
europium	63	**3**	3	4	**3**	3	2	3	3	3
terbium	65	**3**	3	3	**4**	4	2	4	4	3
dysprosium	66	**4**	4	3	**4**	4	2	4	3	4

The sections below provide the detailed assessments for each element. They are informed by the information in Chapter 3 and analysis in Chapter 7, but also take into account other available information impacting material criticality.

Element: Lithium (Li)	**Atomic Number: 3**
Light metallic element with unique electrochemical reactivity properties. Applications include use in ceramics and glass formulations, batteries, lubricating greases, air treatment facilities, continuous casting of metals and primary aluminum production.	
Importance to Clean Energy: Short Term: 2; Medium Term: 3	
The primary clean energy use of lithium is in batteries for electric-drive vehicles. Demand for these batteries-is expected to increase dramatically in the medium term.	
CLEAN ENERGY DEMAND Short Term: 2 Medium Term: 3	• The Li demand for batteries is expected to increase significantly with the deployment of plug-in hybrid electric and all-electric vehicles. In the high material content case, it increases by an order of magnitude. • Batteries could also be important to the successful integration of renewable electricity technologies in the context of Smart Grid development. • In the medium term, Li-ion batteries may gain significant market share in hybrid-electric vehicles, which currently use nickel-metal hydride batteries.
SUBSTITUTABILITY LIMITATIONS *Short Term: 2* *Medium Term: 2*	• Li-ion technology is seen as the most viable option for electric and plug-in hybrid vehicles, but improvements in Li-ion chemistries may significantly enhance the energy density and reduce the specific consumption of Li. • Zinc-air batteries, sodium-sulfur batteries, fuel cells and super- or ultra-capacitors could substitute for Li-ion batteries in stationary configurations. • Substitutes for Li are available for most ceramic, glass and lubricant applications.
Supply Risk: Short Term: 1; Medium Term: 2	
Supply is somewhat constrained by the limits of existing production facilities; a number of options for additional production exist in the medium term. There is no serious indication of long-term physical constraints on supply.	
BASIC AVAILABILITY *Short Term: 1* *Medium Term: 2*	• Produced most economically through saline brine evaporitic processes, but could be produced from hard rock and spodumene deposits in many other countries. • Global production in 2008 was less than 0.5% of global reserves, which are estimated at 10 million tonnes. • Low prices impede increased lithium production, but production has responded to increased demand for lithium used in battery applications. • Recycling has been limited due to dispersion in end-use devices and the high cost of collection, separation and repurification. Successful commercialization of large-scale Li-ion batteries would dramatically improve prospects for recycling.
COMPETING TECHNOLOGY DEMAND *Short Term: 1* *Medium Term: 2*	• Demand in primary aluminum production, ceramics and glass is likely to increase during the next decade, but at a slower rate than Li use for batteries. • Use of Li-ion batteries for “smart-phones,” tablet computers and other hand-held devices could grow rapidly if unit cost reductions increase use.

(Continued)

POLITICAL, REGULATORY AND SOCIAL FACTORS *Short Term: 1* *Medium Term: 1*	• No significant political, regulatory or social factors in the countries producing Li today. • Chile is currently the world's largest producer. • New resources have been discovered in Bolivia, the largest in the world. Political and social factors may inhibit development of Bolivian resources. Bolivian production should not be required to meet forecast demand in the short and medium term.
CO-DEPENDENCE ON OTHER MARKETS *Short Term: 1* *Medium Term: 1*	• No significant co-dependence issues are likely to affect future lithium production.
PRODUCER DIVERSITY *Short Term: 1* *Medium Term: 1*	• Chile, Australia, China and Argentina and the U.S. are all leading mine producers.. • Six countries with major reserves.
References	
USGS (U.S. Geological Survey). 2010. 2008 Minerals Yearbook: Lithium. USGS (U.S. Geological Survey). 2010. Mineral Commodity Summary: Lithium.	
Element: Cobalt	**Atomic Number: 27**
Cobalt is a hard metal used in a wide variety of applications, including high-strength alloys, cutting tools and batteries.	
Importance to Clean Energy: Short Term: 2; Medium Term: 2	
Cobalt is used in several types of batteries for electric vehicles, including nickel metal hydride (NiMH) and lithium nickel cobalt aluminum graphite (Li-NCA-G) (a type of lithium-ion chemistry).	
CLEAN ENERGY DEMAND *Short Term: 2* *Medium Term: 2*	• Li-NCA-G and NiMH batteries contain cobalt. • Battery production for electric drive vehicles could produce a dramatic increase in future cobalt demand.
SUBSTITUTABILITY LIMITATIONS *Short Term: 2* *Medium Term: 2*	• Other lithium-ion battery technologies that do not use cobalt also show promise for use in electric drive vehicles (Gains and Nelson 2010).

(Continued)

Supply Risk: Short Term: 2; Medium Term: 2	
• Cobalt supply has grown significantly over the past decade; new mining projects are scheduled to begin. • Dominance of the Democratic Republic of the Congo (DRC) in cobalt supply is a concern—produces 40% of global cobaltand is consistently ranked below the 10thpercentile in World Governance Indicators (WGI) rankings, including political stability. • Significant new production projected to come on line is from outside of the DRC.	
BASIC AVAILABILITY *Short Term: 1* *Medium Term: 1*	• Production has grown to match demand over the past decade. Production and demand both dropped in 2009 due to the global economic downturn. • Dramatic growth in production is expected to resume and continue through 2014. • Recycling cobalt from lithium ion batteries and metal alloys is economically feasible at current cobalt prices. The European Commission reports that 16% of cobalt use in the EU is from recycled, post-consumer material (EC).
COMPETING TECHNOLOGY DEMAND *Short Term: 2* *Medium Term: 2*	• Catalysts and other chemical products applications have increased demand for cobalt. • Li-Ion and NiMH battery demand has increased dramatically for portable electronics.
POLITICAL, REGULATORY AND SOCIAL FACTORS *Short Term: 3* *Medium Term: 2*	• The DRC dominates cobalt production; the DRC signifies high political risk as it ranks below the 10th percentile in all World Governance Indicators (WGI). DRC dominance to continue in the short term, according to the Cobalt Development Institute. • Higher short term political and social risk factor due to political instability in the DRC. Decreases in the medium term as more non-DNC projects come on line.
CO-DEPENDENCE ON OTHER MARKETS *Short Term: 2* *Medium Term: 2*	• Cobalt can be produced as a primary product or a coproduct. • Most cobalt production is a byproduct from nickel mining or copper mining. • Decreases in nickel or copper demand will negatively affect cobalt supply, but effect may be offset by increased primary cobalt production in the DRC and Morocco.
PRODUCER DIVERSITY *Short Term: 2* *Medium Term: 2*	• Democratic Republic of Congo is dominant producer, but nine other countries have significant production and reserves.
References	
USGS (U.S. Geological Survey). 2010. *2008 Minerals Yearbook*. Reston, VA: USGS. USGS (U.S. Geological Survey). 2010. *Mineral Commodity Summary: Cobalt*. Gains, L. and P. Nelson. 2010. "Lithium-Ion Batteries: Examining Material Demand and Recycling Issues." Argonne NationalLaboratory.	

(Continued)

Element: Gallium (Ga)	**Atomic Number: 31**
A weak metal that is mostly used in electronic devices, primarily in high-speed semiconductors and light-emitting diodes. Other uses include photovoltaic (PV) components, microwave circuitry and infrared technologies.	
Importance to Clean Energy: Short Term: 3; Medium Term: 3	
Uses of gallium in high-efficiency, multi-junction PV cells could increase significantly in the future.	
CLEAN ENERGY DEMAND *Short Term: 3* *Medium Term: 3*	• Demand for high-efficiency PVs is expected to increase throughout the short and medium term. • Growth rate depends on the market success of copper-indium gallium diselenide (CIGS) technology relative to competing PV technologies, as well as on the overall deployment rate of PV systems.
SUBSTITUTABILITY LIMITATIONS *Short Term: 2* *Medium Term: 2*	• CIGS is only one of a number of competing PV technologies, including cadmium telluride andcrystalline silicon.
Supply Risk: Short Term: 1; Medium Term: 1	
Supply risk for gallium is low. Though supply is somewhat constrained by coproduction with aluminum, there are options for additional production in the medium term.	
BASIC AVAILABILITY *Short Term: 1* *Medium Term: 1*	• Appears in trace amounts (<50 parts per million) as a salt in bauxite and zinc ores. Most is extracted from crude aluminum hydroxide solution generated while refining bauxite into aluminum and alumina, the feedstock for aluminum smelters. • Worldwide gallium resources are distributed extensively; no shortage of raw ore. • Primary short-term drivers: rate of global economic recovery, demand for aluminum. • Alumina refining extracts only 10% of Ga in the ore, and only 15% of refiners can recover gallium. Increases in gallium prices could lead to process improvements to increase extraction rates and the installation of gallium recover circuits. • No primary Ga recovery in the United States. • Two U.S. firms produce commercial-grade gallium from impure metal (USGS 2010). • No recycling of Ga from old scrap. • Production of GaAs-based semiconductor devices produces considerable scrap with high-purity GaAs. Much of this "new scrap" is recycled.
COMPETING TECHNOLOGY DEMAND *Short Term: 2* *Medium Term: 3*	• Semiconductor applications requiring high-purity Ga are expected to increase substantially. Short-term demand driven by multi-featured cell and "smart" phones. • Blu-ray video disk players use gallium nitride material. • Other applications: high-concentration PV collectors, large-scale neutrino collectors, biomedical applications, fuel cells and ultra-violet activated phosphor powders.

(Continued)

POLITICAL, REGULATORY AND SOCIAL FACTORS *Short Term: 1* *Medium Term: 1*	• There are no significant political, regulatory or social factors.
CO-DEPENDENCE ON OTHER MARKETS *Short Term: 3* *Medium Term: 3*	• Currently co-produced with aluminum and zinc, but recovery rates could be increased through improved extraction technology.. • Ga does not occur in sufficient concentrations to justify mining solely for its content.
PRODUCER DIVERSITY *Short Term: 1* *Medium Term: 1*	• Primary production in China, Germany, Kazakhstan, Ukraine, Hungary, Japan, Russia and Slovakia. • Significant bauxite deposits in Arkansas, but are not currently economical to produce.
References	
USGS (U.S. Geological Survey). 2009. *2008 Minerals Yearbook: Gallium*. Reston, VA: USGS. USGS (U.S. Geological Survey). 2010. *Mineral Commodity Summary: Gallium*. Reston, VA: USGS.	
Element: Yttrium (Y)	**Atomic Number: 39**
A silvery-metallic metal with chemical properties similar to the lanthanide group. It is a key ingredient in phosphors for both linear fluorescent (LFL) and compact fluorescent (CFL) ligh tbulbs. Yttrium is also used as a red phosphor in televisionsand LCD screens and to increase the strength of aluminum and magnesium structural alloys.	
Importance to Clean Energy: Short Term: 3; Medium Term: 3	
Demand will increase during the switch from current high-volume, halophosphor fluorescent lamps to T8 andT5 linear andCFLs, as a result of DOE rulemaking and worldwide trends. Demand should continue into the medium term until light emitting diode (LED) bulbs achieve significant market penetration.	
CLEAN ENERGY DEMAND *Short Term: 3* *Medium Term: 3*	• U.S. DOE rulemaking will increase demand for T8 and T5 lamps which use Y with terbium to produce “white” light. • U.S. consumer demand for CFLs is growing and the new U.S. federal minimum efficiency standards for general service lighting should dramatically raise CFL demand. • European Union and other regions will implement similar standards to largely eliminate traditional incandescent lamps from the market.
SUBSTITUTABILITY LIMITATIONS *Short Term: 4* *Medium Term: 4*	• No effective substitute for Y as a phosphor in fluorescent lamps has been identified. • Advanced LED technology using greatly reduced or no REEs may begin to replace fluorescent bulbs, but not until well into the medium term. • No known substitutes for Y as a red phosphor in television or LCD screens.
Supply Risk: Short Term: 4; Medium Term: 3	
Not currently mined or refined in the United States; all supplies are imported, predominately from China. Chinese customs tariffs and export quotas could significantly constrain supply/demand balances in the short term. Several possible sources could mitigate this in the medium term.	

(Continued)

BASIC AVAILABILITY *Short Term: 4* *Medium Term: 3*	• China is the largest producer and dominates the market. • Short-term supply will be tight as demand increases faster than Chinese supplies. • Rising prices, export quotas and tariffs coupled with rising domestic internal demand should make development of deposits outside of China more economically feasible. • Significant resources available worldwide in monazite, xenotime ores and ion-absorbing clays. Other producers likely to emerge in the medium term, although supplies will rise only modestly.
COMPETING TECHNOLOGY DEMAND *Short Term: 2* *Medium Term: 2*	• Most used to make phosphors for use in fluorescent lighting, television cathode ray tube displays and LEDs. • Incorporated into electrodes, electrolytes, electronic filters, lasers, superconductors and in advanced medical applications. • The non-phosphor applications are not expected to ramp up drastically.
POLITICAL, REGULATORY AND SOCIAL FACTORS *Short Term: 4* *Medium Term: 3*	• China has instituted significant export quotas and tariffs on all REEs, based chiefly on resource conservation and environmental regulatory reasons. • New mines in Australia, Canada and the United States will provide additional supply, but are subject to strict permitting processes and environmental regulations.
CO-DEPENDENCE ON OTHER MARKETS *Short Term: 2* *Medium Term: 2*	• Found in varying abundance with other REEs, most predominantly in monazite, xenotime ores and ion-absorbing clays. • Most abundant of the heavy rare earth elements, with relatively high revenue stream.
PRODUCER DIVERSITY *Short Term: 4* *Medium Term: 3*	• China currently produces almost all, primarily at Bayan Obo in Inner Mongolia. • China will continue to be the dominant producer in the short and medium term. • New mines in Canada, Australia and the United States will marginally increase world supply in the medium term.
References	
USGS (U.S. Geological Survey). 2010. *Mineral Commodity Summary: Yttrium*. Reston, VA: USGS.	
Element: Indium (IN)	**Atomic Number: 49**
A soft, gray metallic element that is used in indium-tin-oxide (ITO) coatings for highly efficient flat-panel displays. Also usedin infrared detectors, high-speed transistors and in high-efficiency photovoltaic (PV) cells.	
Importance to Clean Energy: Short Term: 3; Medium Term: 3	
Alloyed with copper, gallium and diselenide in ITO coatings for high-efficiency photovoltaic cells.	

(Continued)

CLEAN ENERGY DEMAND *Short Term: 3* *Medium Term: 3*	• Thin-film PV cells utilizing copper-indium-gallium-diselenide (CIGS) alloy.
SUBSTITUTABILITY LIMITATIONS *Short Term: 2* *Medium Term: 2*	• Amorphous silicon or cadmium telluride thin-film PV devices. • Antimony tin oxide could be an alternative to ITO coatings in liquid crystal display flat-panel displays. • Carbon nanotubes could be an alternative to ITO coatings in PV applications and touch screens, but are not yet in widespread use. • Graphene quantum dots could be an alternative to ITO coatings in PV cells.
Supply Risk: Short Term: 3; Medium Term: 2	
Indium is a byproduct of zinc refining. Zinc demand remains strong but future demand depends on macroeconomic conditions in China, Canada, Korea and Japan. There are no significant options for additional primary production in the short term and no refined In was produced in the United States during 2008 or 2009.	
BASIC AVAILABILITY *Short Term: 4* *Medium Term: 3*	• Global demand exceeded production in 2008 and 2009, expected to exceed in 2010. • Market in disequilibrium; production levels have not yet responded to price increases. • Current extraction is inefficient. Increased recovery from tailings could expand production dramatically in the medium term. • Indium has been extensively recovered from the sputtering process for ITO coatings. • Could be recycled from PV modules and flat-panel displays; currently no facilities. • Sub-economic concentrations occur in some copper, lead and tin ores.
COMPETING TECHNOLOGY DEMAND *Short Term: 3* *Medium Term: 4*	• 50% of demand is associated with coatings for flat-panel displays. Retail consumer demand for large, thin, flat-panel displays is likely to increase short-term demand. • Touch-screen displays coatings for point-of-sale retail systems, "smart phones," and tablet computers. Each of these demands is expected to increase in the short term. • Alloying element for the III–V class semiconductors in LEDs and laser diodes. • Emerging uses: electrode-less lamps, mercury alloy replacements and control rods for nuclear power plants. • Each principal demand is growing faster than the rate of general price inflation.
POLITICAL, REGULATORY AND SOCIAL FACTORS *Short Term: 1* *Medium Term: 1*	• No significant political, regulatory or social factors affect indium production.

(Continued)

CO-DEPENDENCE ON OTHER MARKETS *Short Term: 3* *Medium Term: 3*	• In does not occur in concentrations that would justify dedicated mining. • Produced only as a coproduct of zinc mining • Future growth in supply could be limited by slower demand growth for zinc.
PRODUCER DIVERSITY *Short Term: 1* *Medium Term: 1*	• China is dominant producer, but mines in Canada, Japan, Korea, Peru, Belgium and Russia also produce and could expand capacity.
References	
Indium Corporation. 2010. Indium, Gallium and Germanium: Supply and Price Outlook. (Briefing dated April 7). USGS (U.S. Geological Survey). 2009. *2008 Minerals Yearbook: Indium*. Reston, VA: USGS. USGS (U.S. Geological Survey). 2010. *Mineral Commodity Summary: Indium*. Reston, VA: USGS.	
Element: Tellurium (Te)	**Atomic Number: 52**
A brittle, silvery-white metallic element used in photovoltaic (PV) film, steel alloys, rubber processing, synthetic fibers and electronics.	
Importance to Clean Energy: Short Term: 3; Medium Term: 3	
PV films are currently a significant part of global Te demand, mainly due to the rapid expansion of a single company, First Solar. Other PV technologies are available.	
CLEAN ENERGY DEMAND *Short Term: 3* *Medium Term: 3*	• Used in cadmium telluride (CdTe) PV thin films. As of 2008, CdTe was about 8% of the 7 gigawatt (GW) global PV market (DOE 2010) and expanding. • Current demand for PVs accounts for about 17 % of the overall global Te demand. • As the PV market expands, CdTe will likely compete with other PV technologies. • PV industry trend towards reducing material intensity for thin film active layers.
SUBSTITUTABILITY LIMITATIONS	• CdTe is one of a number of PV thin film technologies, including copper indium diselenide (CIS), amorphous silicon and copper zinc tin sulfide (CZTS). Note that CIS also uses indium.
Short Term: 2 *Medium Term: 2*	• Future demand depends on market success of CdTe versus competing PV technologies, as well as the overall deployment rate of PV.
Supply Risk: Short Term: 2; Medium Term: 2	
Te is only produced as a secondary product of copper and to a lesser extent, other nonferrous metals. Though there is onlyone firm in the United States producing commercial-grade tellurium, production is well distributed globally.	

(Continued)

BASIC AVAILABILITY *Short Term: 3* *Medium Term: 3*	• Currently dependent on production of copper. • Production has not increased with production of CdTe PV films and increases in electronic applications, contributing to a constrained supply. This is expected to ease slightly as more Te is extracted from existing anode slimes. • Historically, Te has been recycled from copier drums. Reduced use of these drums has reduced available scrap, meaning currently there is little to no recycling of Te.
COMPETING TECHNOLOGY DEMAND *Short Term: 2* *Medium Term: 2*	• There is some flexibility in the overall demand picture, with the bulk of current Te use currently coming in relatively low-value steel alloys that have alternate formulations. • Recent reductions in use in steel alloys have not quite counterbalanced increases in demand for PV, thermal imaging, thermoelectric applications and other electronics.
POLITICAL, REGULATORY AND SOCIAL FACTORS *Short Term: 1* *Medium Term: 1*	• There are no significant political, regulatory or social factors.
CO-DEPENDENCE ON OTHER MARKETS *Short Term: 3* *Medium Term: 3*	• Co-produced from the anode slimes from electrolytic refining of copper; does not occur in concentrations high enough to justify mining solely for its content. • The price of Te is not high enough to drive increases in copper production, though primary copper production continues to increase globally. • Additional production and recovery methods could mitigate coproduction risk.
PRODUCER DIVERSITY *Short Term: 1* *Medium Term: 1*	• High level of producer diversity; available from the United States, Canada, Japan, Peru, Australia, Belgium, China, Germany, Kazakhstan, the Philippines and Russia.
References	
DOE (U.S. Department of Energy). 2010. 2008 Solar Technologies Market Report. USGS (U.S. Geological Survey). 2009. 2008 Minerals Yearbook: Selenium and Tellurium. Reston, VA: USGS. USGS (U.S. Geological Survey). 2010a. Mineral Commodity Summary. Reston, VA: USGS. USGS (U.S. Geological Survey). 2010b. 2008 Minerals Yearbook: Copper. Reston, VA: USGS.	
Element: Lanthanum (La)	**Atomic Number: 57**
The lightest rare earth element (REE) in the lanthanide series. It is a soft, silvery-white mineral found chiefly in monazite and bastnasite ores.	
Importance to Clean Energy: Short Term: 3; Medium Term: 2	
Importance to clean energy is mainly through battery alloys and phosphors.	

(Continued)

CLEAN ENERGY DEMAND *Short Term: 3* *Medium Term: 2*	• Used in NiMH batteries either as high-purity material or part of mischmetal (a combination of Ce, La, Nd and Pr). NiMH batteries are currently used in almost all hybrid-electric vehicles. • Less than 10% of La supplies are used for lighting phosphors. Demand for phosphors is expected to increase in the short term as fluorescent bulbs gain market share.
SUBSTITUTABILITY LIMITATIONS *Short Term: 3* *Medium Term: 3*	• NiMH batteries already substitute mischmetal compound for pure La. • Li-ion batteries are projected to gain market share; could make up the majority of the HEV battery market by 2020. • There are no substitutes for La as a lighting phosphor in fluorescent light bulbs. • Light-emitting diode (LED) technologies that contain less (or even no) REEs will grow in market share relative to fluorescent light bulbs in the medium term.
Supply Risk: Short Term: 2; Medium Term: 2	
Relatively abundant, though increased clean energy demand and coproduction issues may compromise La supplies in the short and medium terms.	
BASIC AVAILABILITY *Short Term: 2* *Medium Term: 2*	• La is the second most abundant lanthanide after cerium • Relative low value compared to other light and medium REEs with which it is co-produced limits La production, resulting in supply shortages. • Substitute lighting and battery technologies could mitigate demand.
COMPETING TECHNOLOGY DEMAND *Short Term: 2* *Medium Term: 2*	• Other major applications for lanthanum include its use in alloys and fluid cracking catalysts for petroleum refining. Demand growth in these technologies is not expected to be as significant as the demand growth for clean energy applications. • Also used in hydrogen technology applications for hydrogen gas storage and energy conservation. These technologies are not expected to be commercialized and deployed in large numbers until the end of the medium term.
POLITICAL, REGULATORY AND SOCIAL FACTORS *Short Term: 3* *Medium Term: 3*	• Produced predominantly in China, which instituted significant export quotas and tariffs on allREEs for resource conservation and environmental regulatory reasons. • New mines in Australia, Canada and the United States will provide additional supply, but are subject to strict permitting processes and environmental regulations, which have the potential to delay production.
CO-DEPENDENCE ON OTHER MARKETS *Short Term: 2* *Medium Term: 2*	• Second most abundant of all REEs, with a moderately high revenue stream. • Revenue stream has increased significantly relative to other REEs in 2010 due to disproportionate increase in prices for light rare earths.
PRODUCER DIVERSITY *Short Term: 3* *Medium Term: 2*	• Current La production centered in China. • The United States has one domestic source (Molycorp) and other potential future domestic sources. • Large amounts of La exist in monazite ores found in India, Brazil, Australia and Africa. • By 2015, non-Chinese mines are expected to provide significant additional production.

(Continued)

References	
Molycorp Minerals. 2009. "Lanthanum." http://www.molycorp.com/lanthanum.asp. Oakdene Hollins. 2010. "Lanthanide Resources and Alternatives." Aylesbury, UK: Department for Transport and Department for Business, Innovation and Skills. http://www.oakdenehollins.co.uk/.	
Element: Cerium (Ce)	**Atomic Number: 58**
Ce is a ductile and malleable light rare earth element (LREE) with atomic number 58.	
Importance to Clean Energy: Short Term: 3; Medium Term: 2	
Used in nickel metal hydride (NiMH) batteries found in most hybrid and electric vehicles and in phosphor powders in linearfluorescent and compact fluorescent light bulbs.	
CLEAN ENERGY DEMAND *Short Term: 3* *Medium Term: 2*	• Demand is expected to increase in the short term with demand for phosphors and hybrid vehicles. • Demand growth may decrease in the medium term as lithium-ion (Li-ion) technology improves and Li-ion batteries supplant NiMH in hybrids. • Despite the falling market share in the medium term, the overall market for NiMH may continue to rise as the number of hybrids manufactured each year increases.
SUBSTITUTABILITY LIMITATIONS	• Limited substitutability within phosphors and NiMH batteries. • Li-ion batteries are projected to gain market share; could make up the majority of the HEV battery market by 2020. •
Short Term: 2 *Medium Term: 2*	• Light emitting diodes (LED) use little or no REEs; could replace fluorescent light bulbs. LEDs for room lighting are not expected to be cost competitive until the medium term.
Supply Risk: Short Term: 2; Medium Term: 2	
New Ce supply sources from outside China are projected to meet future demand growth. Ce likely to be in surplus both in the short and the medium term.	
BASIC AVAILABILITY *Short Term: 1* *Medium Term: 1*	• Supply adequate in the short and medium term. • Future supply growth hinges on new mining projects; Ce is the most abundant REE and will be produced by all new mines.
COMPETING TECHNOLOGY DEMAND *Short Term: 2* *Medium Term: 3*	• Other applications including polishing compounds, catalysts, fuel additives, glass and enamel additives, permanent magnets. • Emerging use in nano technologies, but not likely to be commercialized until at least the medium term.
POLITICAL, REGULATORY AND SOCIAL FACTORS *Short Term: 3* *Medium Term: 3*	• Current production centered in China, which has significant REE export quotas and tariffs. • New mines in Australia, Canada and the United States will increase supply, but are subject tostrict permitting and environmental regulations.

(Continued)

CO-DEPENDENCE ON OTHER MARKETS *Short Term: 2* *Medium Term: 2*	• Most abundant of all REEs. • High revenue stream despite relatively low price. • Revenue stream has increased dramatically in response to export restrictions that disproportionately increased prices for LREEs. Highest revenue stream of any REE in the second half of 2010.
PRODUCER DIVERSITY *Short Term: 3* *Medium Term: 2*	• China is currently the dominant producer. • All new non-Chinese mines will produce significant amounts of cerium, increasing diversity more than for most other rare earth elements.
References	
AZoMaterials. 2001. "Cerium." Accessed July 20. http://www.azom.com/details.asp?ArticleID=592. Oakdene Hollins. 2010. "Lanthanide Resources and Alternatives." Aylesbury, UK: Department for Transport and Department for Business, Innovation and Skills. http://www.oakdenehollins.co.uk/ GE (General Electric). 2010. "Response to Department of Energy Request for Information."	
Element: Praseodymium (Pr)	**Atomic Number: 59**
A light rare earth element (LREE) used in a variety of technologies, including clean energy. It is a soft, silvery, malleable and ductile metal. Pr is paramagnetic at any temperature above 1 K.	
Importance to Clean Energy: Short Term: 2; Medium Term: 2	
Can partially substitute for neodymium in neodymium-iron-boron magnets for electric vehicle motors and wind turbine generators. Pr and several other LREEs are also used in mischmetal for nickel metal hydride batteries.	
CLEAN ENERGY DEMAND *Short Term: 2* *Medium Term: 2*	• Praseodymium is used as a minor constituent of neodymium magnets and mischmetal in nickel metal hydride batteries. • Substituted for neodymium (Nd) in NdFeB permanent magnets at a ratio of 1 Pr for 4 Nd. This ratio matches the natural abundance of the elements, reducing the Nd requirement while increasing the limited market for Pr (Ames Laboratory 2010). • May help augment the field strength of NdFeB magnets when it is reduced by the addition ofdysprosium (to increase temperature performance) (Hykaway 2010).
SUBSTITUTABILITY LIMITATIONS *Short Term: 1* *Medium Term: 1*	• Generally used as a substitute for other REEs in clean energy applications, not as a primary material.
Supply Risk: Short Term: 2; Medium Term: 2	
Although it is the least abundant of the light REEs, Pr supply should meet demand in the short and medium term.	

(Continued)

BASIC AVAILABILITY *Short Term: 2* *Medium Term: 2*	• Least abundant LREE in bastnasite and monazite, but supply projected to meet demand in the short and medium term. • New mines will significantly increase supply by 2015 (see table 7-2).
COMPETING TECHNOLOGY DEMAND *Short Term: 1* *Medium Term: 2*	• Used in lighter flints, glass polishing, lasers, magnets and batteries. • Emerging applications include magnetic refrigeration, high-temperature superconductivity and hydrogen storage. • Demand for Pr in clean energy applications is expected to increase more significantly than in non-clean energy applications.
POLITICAL, REGULATORY AND SOCIAL FACTORS	• Produced predominantly in China, which instituted significant export quotas and tariffs on REEs for resource conservation and environmental regulatory reasons.
Short Term: 3 *Medium Term: 3*	• New mines in Australia, Canada and the United States will provide additional supply, but are subject to strict permitting processes and environmental regulations.
CO-DEPENDENCE ON OTHER MARKETS *Short Term: 3* *Medium Term: 3*	• Less abundant than other LREEs in bastnazite or monazite. • Moderate revenue stream is lower than other LREEs but comparable with HREEs. • Revenue stream has increased less than other LREEs in response to recent export restrictions.
PRODUCER DIVERSITY *Short Term: 3* *Medium Term: 2*	• Additional short-term sources may include Molycorp, Mount Weld and Nolans. • New mines projected to increase supply by almost 50% in the short term (see figure 7-2). • Additional medium-term production capacities and producer diversity is expected.
References	
Ames Laboratory. 2010. "The Ames Laboratory Response to U.S. Department of Energy's Request for Information." Hykaway, J. 2010. *The Rare Earths: Pick Your Spots Carefully*. Byron Capital Markets.	
Element: Neodymium (Nd)	**Atomic Number: 60**
A light rare earth element (LREE) used in high-strength permanent magnets. Other applications include use as a component of didymium for coloring glass and ceramics, astronomical instruments and glass lasers.	
Importance to Clean Energy: Short Term: 3; Medium Term: 4	
Primarily used in clean energy technologies as an alloy with iron (Fe) and boron (B) to form high-strength Nd-Fe-B magnets used extensively in high-efficiency, brushless motors in electric vehicles and in direct-drive generators. Applications includewind turbines; hybrid, plug-in hybrid and all-electric vehicles; and energy efficient appliances.	

(Continued)

CLEAN ENERGY DEMAND *Short Term: 3* *Medium Term: 4*	• Demand for Nd in magnets depends on the global economic recovery and the success of "Green Economy" efforts in the United States, European Union, Japan and China. • Nd is also a component of mischmetal used in nickel metal hydride batteries. • 80% of global consumption of neodymium oxide (Nd2O3) in 2009 was for high-strength magnet applications; only 10% were for wind generators and hybrid vehicles. • Nd use in clean energy is expected to grow with increasing market penetration of electric drive vehicles and permanent magnet wind turbines.
SUBSTITUTABILITY LIMITATIONS	• Limited substitutes for Nd-Fe-B magnets.
Short Term: 3 *Medium Term: 3*	• Rare earth magnets constructed from an alloy of samarium and cobalt could work as a substitute, but these are generally more expensive than Nd-Fe-B magnets. • At the component level, there are substitute motor and generator technologies that do not use rare earth elements.
Supply Risk: Short Term: 4; Medium Term: 3	
Increased demand for Nd will lead to tight supplies in the short term. Additional mines outside of China may be brought into commercial production, reducing the potential for supply shortages in the medium term.	
BASIC AVAILABILITY *Short Term: 4* *Medium Term: 3*	• Significant price increase for Nd in last year; supply little increased from 2008 level. • Limited near-term flexibility for increasing global supply, despite stockpiled supplies. • Demand for Nd-Fe-B magnets is likely to exceed producers' ability in the short term. • Nd is not recycled. Recycling from magnets in electric drive vehicles and wind turbines could become economical in the medium term. Recycling Nd-Fe-B magnets from consumer electronic devices is unlikely to be economic in the United States in the short or medium term.
COMPETING TECHNOLOGY DEMAND *Short Term: 2* *Medium Term: 2*	• Demand for Nd-Fe-B magnets is expected to grow faster than overall U.S. economy. • There is no large emerging competing demand for Nd. • Magnetic refrigeration and permanent magnet motors for home appliances could significantly increase demand for Nd-Fe-B magnets beyond the medium term.
POLITICAL, REGULATORY AND SOCIAL FACTORS *Short Term: 4* *Medium Term: 4*	• Predominantly produced in China, which has instituted significant export quotas and tariffs on REEs for resource conservation and environmental regulatory reasons. • New mines in Australia, Canada and the United States will provide additional supply, but are subject to strict permitting processes and environmental regulations.
CO-DEPENDENCE ON OTHER MARKETS *Short Term: 2* *Medium Term: 2*	• Moderate abundance and prices compared to other REEs lead to high revenue streams. • Neodymium usually drives production of other REEs.

(Continued)

PRODUCER DIVERSITY *Short Term: 4* *Medium Term:3*	• Mainly produced from mines in China. • New non-Chinese mines will increase diversity significantly by 2015, even though global supply projected to remain tight (see Table 7-2).
References	
GE (General Electric). 2010. Submission in Response to the U.S. Department of Energy's Request for Information. USGS (U.S. Geological Survey). 2009. *2008 Minerals Yearbook: Rare Earths*.	
Element: Samarium (Sm)	**Atomic Number: 62**
A light rare earth element (LREE) used in a number of applications, including magnets, military equipment, catalysts and nuclear reactors. It is a lustrous silver-white metal found along with other REEs in monazite, bastnasite and samarskite geological deposits.	
Importance to Clean Energy: Short Term: 1; Medium Term: 1	
Samarium cobalt (SmCo) permanent magnets are slightly less powerful by size and weight than non-Sm containing neodymium-iron-boron (NdFeB) magnets, though SmCo permanent magnets have higher temperature ratings that make them appropriate for certain motor and generator applications (Electron Energy Corporation 2010). SmCo permanent magnets are not currently used extensively in clean energy applications.	
CLEAN ENERGY DEMAND *Short Term: 1* *Medium Term: 1*	• SmCo magnets are not likely to be used extensively in clean energy applications in the short or medium term.
SUBSTITUTABILITY LIMITATIONS *Short Term: 1* *Medium Term: 1*	• Due to the limited use of samarium in clean energy applications, substitutability is not an issue.
Supply Risk: Short Term: 2; Medium Term: 2	
Samarium is projected to be in excess supply both in the short and medium term.	
BASIC AVAILABILITY *Short Term: 2* *Medium Term: 2*	• 2010 production is projected to be approximately twice current demand (Kingsnorth 2010). • This trend is forecast to continue, with new mines opening by 2015 projected to increase production capacity by about 30% (see Table 7.2).
COMPETING TECHNOLOGY DEMAND *Short Term: 1* *Medium Term: 2*	• Used to manufacture components for industrial, commercial and military uses. • SmCo magnets are used in precision-guided weapons due to their ability to operate at high temperatures. • Samarium oxide is used as a neutron absorber in nuclear power plants.

(Continued)

POLITICAL, REGULATORY AND SOCIAL FACTORS *Short Term: 3* *Medium Term: 3*	• Produced predominantly in China, which instituted significant export quotas and tariffs on REEs based on resource conservation and environmental regulations. • New mines in Australia, Canada and the United States will provide additional supply, but are subject to strict permitting processes and environmental regulations, which have the potential to delay production.
CO-DEPENDENCE ON OTHER MARKETS *Short Term: 3* *Medium Term: 3*	• Moderately abundant compared to other REEs. • Limited demand and correspondingly low relative prices mean that other more valuable REEsare more likely to drive production decisions.
PRODUCER DIVERSITY *Short Term: 3* *Medium Term: 2*	• Samarium is found in significant quantities in non-Chinese mines likely to begin production inthe short to medium term. Samarium production from non-Chinese sources is likely to account for over 25% of global supply by 2015 (see Table 7-2).
References	
Kingsnorth, D. 2010. Interview. October 10. Electron Energy Corporation. 2010. "Response to Department of Energy Request for Information." June 7.	
Element: Europium (Eu)	**Atomic Number: 63**
A heavy rare earth element (HREE) with the atomic number 63, Eu is a ductile metal with the same relative hardness of lead. Combining Eu phosphor compounds with terbium phosphor compounds produces the white light of helical fluorescent light bulbs and is a primary component in the production of T8 and T5 fluorescent tubes.	
Importance to Clean Energy: Short Term: 3; Medium Term: 3	
Demand will increase during the anticipated switch from high-volume halophosphor fluorescent lamps to T8 andT5 linear and compact fluorescent tubes as a result of DOE rulemaking and worldwide trends. Increased demand is expected until light emitting diode (LED) bulbs (which use much less REEs) achieve significant market penetration.	
CLEAN ENERGY DEMAND *Short Term: 3* *Medium Term: 3*	• Beginning in July 2012, U.S. DOE rulemaking on general service fluorescent lamps will increase demand for linear fluorescent lamps (LFLs), which use Eu phosphors. • U.S. consumer demand for compact fluorescent lights (CFLs) is growing. On January 1, 2012, new U.S. federal minimum efficiency standards for general service lighting will dramatically raise demand for CFLs and consequently demand for Eu. • Similar standards will be implemented in the European Union and other regions and will largely eliminate traditional incandescent lamps from the market.
SUBSTITUTABILITY LIMITATIONS *Short Term: 4* *Medium Term: 4*	• No proven substitute for Eu in fluorescent lamps has been identified. • No known substitutes for Eu as a red phosphor in television or LCD screens. • Advanced LED technology using greatly reduced or no REEs may begin to replace fluorescent bulbs, but not until well into the medium term.

(Continued)

Supply Risk: Short Term: 3; Medium Term: 3	
As with most REEs, and especially with the “heavy” elements like Eu, the majority of current supply comes from China. Industry expert predicts a supply shortage as early as 2011. Several possible sources could mitigate constriction in the medium term.	
BASIC AVAILABILITY *Short Term: 3* *Medium Term: 3*	• Eu is one of the scarcest REEs. • Demand for Eu is expected to exceed supply in the 2014 and 2015 timeframe (Oakdene Hollins 2010). • In the medium term, new mines projected to significantly increase supply (see Table 7-2) .although demand will continue to grow with increased use of LFLs and CFLs.
COMPETING TECHNOLOGY DEMAND *Short Term: 2* *Medium Term: 2*	• As an activator for yttrium-based phosphors in television and LCD screens. • Expanded use in nuclear reactors, due to great affinity to absorb neutrons. • Eu is also used to dope glasses and plastics for laser production, to investigate bio-molecular reactions during drug screening trials and as a counterfeiting indicator on Euro banknotes. • No competing use is expected to increase demand as rapidly as lighting.
POLITICAL, REGULATORY AND SOCIAL FACTORS *Short Term: 4* *Medium Term: 3*	• Predominantly produced in China, which instituted significant export quotas and tariffs on REEs based on resource conservation and environmental regulations. • New mines in Australia, Canada and the United States will provide additional supply, but are subject to strict permitting processes and environmental regulations.
CO-DEPENDENCE ON OTHER MARKETS *Short Term: 3* *Medium Term: 3*	• Eu supply from China occurs as a byproduct of the yttrium-rich, ion-adsorption clay ores in the south China region and in bastnasite ores from Mongolia. Both yttrium and Eu are in highdemand; co-dependence should diminish the supply risk of each.
PRODUCER DIVERSITY *Short Term: 4* *Medium Term: 3*	• Although China currently produces almost all Eu, non-Chinese mines coming on line by 2015 are expected to significantly increase the diversity of supply (see Table 7-2).
References	
Oakdene Hollins. 2010. “LanthanideResources and Alternatives.”Aylesbury, UK: Department for Transport and Department for Business, Innovation and Skills. http://www.oakdenehollins.co.uk/.	
Element: Terbium (Tb)	**Atomic Number: 65**
A heavy rare earth element used in fluorescent lighting phosphors and magnets for electric motors.	
Importance to Clean Energy: Short Term: 3; Medium Term: 3	
Used in several clean energy applications, most notably as a phosphor in fluorescent light bulbs. Can also be used instead of dysprosium (Dy) as an additive in Nd-Fe-B permanent magnets.	

(Continued)

CLEAN ENERGY DEMAND *Short Term: 3* *Medium Term: 3*	• Replacement of incandescent with fluorescent lighting expected to increase consumption in the short and medium terms. • Light emitting diode (LED) lighting could reduce long-term demand.
SUBSTITUTABILITY LIMITATIONS *Short Term: 3* *Medium Term: 3*	• No current substitutes for Tb as a lighting phosphor in fluorescent bulbs. Ongoing research, particularly in Japan, seeks to reduce Tb required in phosphors (General Electric 2010). • Advanced LED bulbs that do not use Tb may replace fluorescent lighting, but not until at leastthe medium term. • Tb and Dy can both be used as additives in Nd-Fe-B magnets. Tb is historically more expensive than Dy, so its use in magnets is already limited.
Supply Risk: Short Term: 4; Medium Term: 4	
In the short term, Tb supply will be very tight relative to demand. In the medium term, additional non-Chinese producers will increase supply, but may not compensate for reduced Chinese supply and increased world demand.	
BASIC AVAILABILITY *Short Term: 4* *Medium Term: 4*	• Demand is expected to exceed global mine production in 2010. • Western mines will increase supply in the short term, but supply will remain tight throughoutthe medium term due to increased demand and decreased Chinese production. • Not currently recycled. Recycling from fluorescent light bulbs possible, if economic.
COMPETING TECHNOLOGY DEMAND *Short Term: 2* *Medium Term: 2*	• Tb also has properties that make it suitable for use in magnetic refrigeration or as a stabilizer in fuel cells. • Competing technology demand not likely to grow as fast as clean energy demand.
POLITICAL, REGULATORY AND SOCIAL FACTORS *Short Term: 4* *Medium Term: 4*	• Predominantly produced in China, which instituted significant export quotas and tariffs on REEs based on resource conservation and environmental regulations. • New mines in Australia, Canada and the United States will provide limited additional supply and are subject to strict permitting processes and environmental regulations.
CO-DEPENDENCE ON OTHER MARKETS *Short Term: 4* *Medium Term: 4*	• Co-produced from REE ores along with other lanthanides. • Tb scarce even compared to other REEs—most deposits contain Tb concentrations of less than 1% by weight of total REO (Kingsnorth 2010). • Low revenue stream even with high prices. Revenue stream has remained relatively flat while LREE revenues have increased due to recent export restrictions.

(Continued)

PRODUCER DIVERSITY *Short Term: 4* *Medium Term: 3*	• Western mines will provide significant percentage of supply by 2015 (see Table 7-2). • China will remain the dominant producer in the medium term.
References	
GE (General Electric). 2010. "Response to U.S. Department of Energy's Request for Information." June 7. Kingsnorth, D. 2010. "Rare Earths: Facing New Challenges in the New Decade." Presented at the Society for Mining Metallurgy and Exploration Annual Meeting. Phoenix.	
Element: Dysprosium (Dy)	**Atomic Number: 66**
Dy is a heavy rare earth element. It is a soft metal with a silver luster and extremely high magnetic strength. Primary uses included ceramics, high-intensity lighting and as an additive to rare earth permanent magnets.	
Importance to Clean Energy: Short Term: 4; Medium Term: 4	
Dysprosium's primary clean energy use is as an additive to neodymium-iron-boron (NdFeB) magnets. The addition of either Dy or terbium (at approximately 5% of the magnet's weight) helps raise the increase the "Curie temperature" at which the magnet can operate before losing its magnetic field.	
CLEAN ENERGY DEMAND *Short Term: 4* *Medium Term: 4*	• Although used in relatively small quantities in magnets, it is crucial for magnets capable of high-temperature operations (particularly in vehicle drives). • Demand for Dy will increase significantly with the growing market for electric drive vehicles in both the short and medium term.
SUBSTITUTABILITY LIMITATIONS	• The only known substitute in permanent magnets is terbium, which is even rarer and historically more expensive.
Short Term: 3 *Medium Term: 3*	• At the component level, there are substitute motor and generator technologies that do not use rare earth elements.
Supply Risk: Short Term: 4; Medium Term: 4	
More than 90% of the global supply of Dy comes from China. Dependence on Chinese exports is expected to lead to a critical shortage of the element between 2012 and 2014. New mines are scheduled to come online in the medium term that could mitigate this constriction.	
BASIC AVAILABILITY *Short Term: 4* *Medium Term: 4*	• Demand projected to increase significantly with minimal increased supply (Oakdene Hollins 2010). • By 2015, global production capacity is expected to increase by less than 15% (see Figure 7-2).
COMPETING TECHNOLOGY DEMAND *Short Term: 2* *Medium Term: 2*	• Emerging technologies exist but unlikely to significantly increase demand pressure compared to current applications. • Expanded uses could include magneto-mechanical sensors, actuators and acoustic and ultrasonic transducers, e.g. flat-panel speakers. Dy has been considered for use in diesel engine fuel injectors.

(Continued)

POLITICAL, REGULATORY AND SOCIAL FACTORS *Short Term: 4* *Medium Term: 4*	• China has instituted significant export quotas and tariffs on all REEs based on resource conservation and environmental regulatory reasons. • New mines in Australia, Canada and the United States will provide little additional supply and are subject to strict permitting processes and environmental regulations, which have the potential to delay production.
CO-DEPENDENCE ON OTHER MARKETS *Short Term: 3* *Medium Term: 3*	• Moderate revenue stream due to low abundance and historically high prices. • Revenue stream has remained flat compared to LREEs, whose prices have increased dramatically due to recent export restrictions. • Most abundant of the HREEs in many Chinese ores, but found in low concentrations elsewhere.
PRODUCER DIVERSITY *Short Term: 4* *Medium Term: 4*	• Found in relatively small amounts in new Western mines scheduled to begin production in the short and medium term. Mountain Pass will produce minimal amounts of dysprosium, and other new mines scheduled to begin production by 2015 will increase production by less than 15% over current levels (see Figure 7-2).
References	
GE (General Electric). 2010. "Response to U.S. Department of Energy's Request for Information." June 7. Kingsnorth, D. 2010. "Rare Earths: Facing New Challenges in the New Decade." Presented at the Society for Mining Metallurgy and Exploration Annual Meeting. Phoenix. Oakdene Hollins. 2010. "Lanthanide Resources and Alternatives." Aylesbury, UK: Department for Transport and Department for Business, Innovation and Skills. http://www.oakdenehollins.co.uk/.	

Appendix B: Market Share Assumptions and Material Content Calculation

Batteries in Electric Drive Vehicles

Projections for key material use in batteries consider two different battery types, nickel metal hydride (NiMH) and lithium-ion (Li-ion). Three different types of electric drive vehicles were considered: hybrid electric (HEV), plug-in hybrid electric (PHEV) and electric (EV).

Market Share

Market share assumptions for batteries used in each type of vehicle are:

- **Hybrid-electric:** All HEVs are assumed to use NiMH batteries.
  Rationale: The 100% NiMH market share assumption contributes to the highest probable estimate of requirements for NiMH battery materials. NiMH battery technology is relatively safe, mature and cost effective for the storage requirements of hybrids. Some experts predict that lithium-ion battery technology will capture an increasing share of the HEV market in the medium term. However, the pace at which the shift will occur is uncertain, and it will require lithium-ion batteries to become significantly more cost competitive (Oakdene Hollins 2010).
- **Plug-in Hybrid and Electric**: All PHEVs and EVs are assumed to use lithium-ion batteries.
  Rationale: Although there are still significant price, safety and reliability issues with lithium- ion batteries, they are considered the only batteries likely to meet the weight and performance requirements of PHEVs and EVs in the short and medium term (National Research Council 2010). The first mass-market PHEVs and EVs, including the Chevy Volt and Nissan Leaf, will all use lithium-ion batteries.

Material Intensity: NiMH batteries

The material intensity of elements used in NiMH batteries was calculated based on several assumptions about battery capacity and chemistry (i.e, anode and cathode composition) for a battery with power rating and cell voltage equivalent to the battery used in a third-generation Toyota Prius.

- **Positive electrode capacity:** A total power rating of 1.3 kWh and 1.2 V/cell was assumed. This yields a total *positive electrode capacity* of 1,083 Ah [=1.3kWh/1.2V*1000].
- **Negative electrode capacity:** This is calculated as negative electrode capacity = (positive electrode capacity)*n/p, where *n/p* is the assumed ratio of negative to positive electrode capacity. For the high material content case, an n/p ratio of 1.8 is assumed, which represents a likely value for current generation technology. For low material content, an n/p ratio of 1.2 is assumed, which represents a lower value that is technically feasible. These assumptions yield *negative electrode capacity* values of 1,950 Ah (high material content) and 1,300 Ah (low material content).

- **Weight of negative electrode alloy:** This is calculated based on an assumption of 300 Ah/kg of alloy. This yields a high *negative electrode alloy weight* of 6.5 kg alloy and a low *negative electrode alloy weight* of 4.3 kg.
- **Negative electrode alloy composition:** The battery alloy is assumed tobe AB5 employing a widely used composition of mischmetal: $La_{5.7}Ce_{8.0}Pr_{0.8}Nd_{2.3}Ni_{59.2}Co_{12.2}Mn_{6.8}A_{l5.2}$ with a total molar weight of 70.6 g. Based on this formula, the high and low weight content of individual elements is given in Table B-1. NiMH material intensity for batteries used by other manufacturers would likely vary with battery performance specifications and the composition of the battery allow.

Estimates for positive electrode capacity were from DOE EERE (2009). Estimates for the n/p ratio and weight ratio of negative electrode alloy were provided by DOE EERE (2010). Alloy composition is based on Linden's Handbook of Batteries (Reddy 2011).

Material Content: Lithium-Ion Batteries

Material content assumptions for lithium-ion batteries in electric and plug-in hybrid electric vehicles are taken from calculations by Gains and Nelson (2009). The authors identified four lithium-ion battery chemistries with potential for automotive applications. For each of the four chemistries, they estimated total lithium content for vehicles with ranges of 4, 20, 40 and 100 miles. Cobalt content for each combination of battery chemistry and vehicle range was calculated for this Strategy using the authors' results for lithium content and the molecular formula of the battery cathode material.

Battery sizes for PHEVs and EVs are based on ranges suggested by Gains and Nelson (2009). PHEVs are assumed to have a 40-mile range, and EVs a 100-mile range. To determine high and low material content, the highest and lowest values of lithium and cobalt were selected from among the four different battery chemistries.

The battery chemistries and material content values selected are shown in Table B-2.

Table B-1. Material Intensity Calculations for NiMH Batteries

Element	Molar % in AB5	Weight % in AB5	kg per battery high	kg per battery low
La	5.7%	11.2%	0.73	0.49
Ce	8.0%	15.9%	1.03	0.69
Pr	0.8%	1.6%	0.10	0.07
Nd	2.3%	4.7%	0.31	0.20
Ni	59.2%	49.2%	3.20	2.13
Co	12.2%	10.2%	0.66	0.44
Mn	6.8%	5.3%	0.34	0.23
Al	5.2%	2.0%	0.13	0.09
Total			**6.5**	**4.33**

Table B-2. Material Content Calculations for Lithium-ion Batteries

Application	Material	High/ Low	Material content (kg)	Battery Chemistry Designation	Cathode	Anode
PHEV 40	*Lithium*	Low	1.35	LMO-G	$LiMn_2O_4$	Graphite
		High	5.07	LMO-TiO	$LiMn_2O_4$	$Li_4Ti_5O_{12}$
	Cobalt	Low	0	LMO (both)	$LiMn_2O_4$	Either
		High	3.77	NCA-G	$LiNi_{0.8}Co_{0.15}Al_{0.05}O_2$	Graphite
EV 100	*Lithium*	Low	3.38	LMO-G	$LiMn_2O_4$	Graphite
		High	12.68	LMO-TiO	$LiMn_2O_4$	$Li_4Ti_5O_{12}$
	Cobalt	Low	0	LMO (both)	$LiMn_2O_4$	Either
		High	9.41	NCA-G	$LiNi_{0.8}Co_{0.15}Al_{0.05}O_2$	Graphite

Permanent Magnets in Wind Turbines and Electric Drive Vehicles

Market Share

Market share assumptions for wind turbines and electric drive vehicles employing neodymium-ironboron (NdFeB) permanent magnet (PM) motors are:

- **Wind turbines:** The low market share assumption is that 10% of both onshore and offshore wind turbines will use NdFeB PM generators. The high market share assumption is that 25% of onshore wind turbines and 75% of offshore wind turbines will use NdFeB PM generators.
  Rationale: There is no publicly available data on current market share, but anecdotal discussions with industry experts indicates that only a small percentage of the wind turbines currently use rare earth permanent magnets. Therefore, the low market share represents a continuation of current market share trends. The high market share assumption is based on the preference for NdFeB permanent magnet generators in larger wind turbines (in the 2-3+ MW range) and the trend towards the use of larger turbines in new wind projects, particularly for offshore applications.
- **Electric Drive Vehicles:** All electric drive vehicles (HEVs, PHEVs and EVs) are assumed to use NdFeB PM motors.
  Rationale: Due to their superior power to weight ratio, NdFeB PM motors are used in current model HEVs, as well as the Chevy Volt PHEV and Nissan Leaf EV. These PM motors are expected to dominate the electric drive vehicle market well into the medium term, although new motors without rare earth materials are being developed (Aston 2010).

Therefore, the 100% NdFeB PM market share assumption contributes to the highest probable estimate of material requirements.

Material Intensity

Material intensity for NdFeB PM motors and generators is calculated as follows:

- **Wind turbines:** Material intensity for neodymium and dysprosium is calculated from the estimated weight of total NdFeB magnet material per MW of turbine output. High and low estimates for total magnet weight are 600 kg/MW and 400 kg/MW, respectively, based on Arnold Magnetics RFI submission (2010). Neodymium and dysprosium content is estimated to be 31% and 5.5% of magnet weight, respectively, assuming a NdFeB-AH magnet composition (Electron Energy Corporation 2010).
- **Electric Drive Vehicles:** Material intensity for neodymium and dysprosium is calculated from the estimated weight of total NdFeB magnet material per vehicle motor. The high weight estimate for magnets is 2 kg per vehicle, based on General Electric (2010). The low estimate is 1 kg/vehicle, based on Lifton (2009). Neodymium and dysprosium content is estimated to be 31% and 5.5% of magnet weight, respectively, assuming a NdFeB-AH magnet composition (Electron Energy Corporation 2010).

The difference between the high and low material content estimates for magnets accounts for incremental improvements that are likely to occur in the short to medium term, as well as wide variations in individual manufacturers specifications for magnets. Material content estimates also assume that magnets do not contain praseodymium, which can reportedly be substituted for neodymium (up to 25%) to reduce cost and increase corrosion resistance (Oakdene Hollins 2010). The actual extent of this substitution is uncertain.

Photovoltaic (PV) Cells

Market Share

Market share assumptions for cadmium telluride (CdTe) and copper-indium-gallium-diselenide (CIGS) PV cells are similar. For both technologies, high and low market share assumptions are 10% and 50%, respectively, of total added PV capacity.

Rationale: This assumption reflects the likelihood that both of these thin-film PV technologies will continue to mature and capture greater shares of the total PV market, which is currently dominated by conventional silicon based cells. Both are viable and neither currently has a clear advantage over the other (Grana 2010). It is unlikely that market share for both technologies will be near the high end of the market share estimate (50% each for both CIGS and CdTe) at the same time.

Material Intensity

Material intensity calculations for tellurium, gallium and indium used in CdTe and CIGS PV cells are:

- **Tellurium intensity in CdTe:** High materials intensity assumes 3.1 grams tellurium per cubic centimeter (cc) of thin film, a 2.1 micron absorber layer and 10% cell efficiency. This yields a material intensity of 0.145 g/W or 145 tonnes tellurium per GW. Low materials intensity assumes 3.1 g Te/cc of thin film, a 1 micron absorber layer and 14.4% cell efficiency. This yields a material intensity of 43 tonnes tellurium per GW.

- **Gallium intensity in CIGS:** In the High Materials Intensity case, gallium represents 6% of the mass, there is about 0.4 g/cm^3 gallium included in a 2.5 micron absorber layer and cell efficiency is 10%, leading to a materials requirement of 20 tonnes gallium per GW. In the low materials intensity case, the Ga represents about 8% of the mass, gallium density is assumed to be about 0.5 g/cm3, with an active layer of 1.0 micron and a cell efficiency to 14%, leading to a materials requirement of 4 tonnes/GWp.
- **Indium Intensity in CIGS:** In the High Materials Intensity case, indium represents 27% of the mass, there is about 2.0g/cm^3 indium included in a 2.5 micron absorber layer and the cell efficiency is 10%. This leads to a materials requirement of 110 tonnes indium per GW. In the low materials intensity case, the indium represents about 30% of the mass; indium density is assumed to be about 2.2 g/cm3, with an active layer of 1.0 micron; and cell efficiency is 14%. This leads to a materials requirement of 16.5 tonnes indium per GW.

All assumptions used in the calculations were provided by NREL (2010).

Lighting Phosphors

Due to a lack of available data on the deployment of high efficiency light bulbs, demand projections for phosphors were not developed using the same type of market share or material intensity calculations developed for batteries, magnets or PV cells. Instead, projections are based on high and low rates of compound growth from estimated 2010 lighting phosphor demand, combined with assumptions about average weight percentages of each rare earth used in phosphors.

Estimated total phosphor demand in 2010 is reported by Lynas Corporation as 7,900 tonnes rare earth oxide (REO). Based on estimates from a number of lighting industry experts, eighty-five percent of this phosphor demand is assumed to be for lighting, yielding a total 2010 lighting phosphor demand of 6,715 tonnes REO.

Lighting phosphor demand for individual rare earth elements in 2010 was calculated based on estimated percentage breakdowns of total phosphor demand by element provided by Lynas (2010). The results are shown in table B-3.

Table B-3. Demand Calculations for Lighting Phosphors in 2010

Element	Percentage of Total Phosphor Demand (source: Lynas 2010)	2010 Demand (Tonnes REO)
Lanthanum	8.5%	571
Cerium	11.0%	739
Europium	4.9%	329
Terbium	4.6%	309
Yttrium	69.2%	4647

High and low estimates of annual lighting phosphor demand by element are calculated from the 2010 demand values by assuming a low annual growth rate of 2.2% and high growth rate of 3.5%. Growth rate assumptions are based on the growth rates for compact fluorescent lamps (CFLs) in the IEA (2010) "Phase Out of Incandescent Lamps" information paper.

APPENDIX C. 111TH CONGRESS RARE EARTHS AND CRITICAL MATERIALS LEGISLATION

Legislation	House	Senate	House
Title	Rare Earth Supply-Chain Technology and Resources Transformation Act of 2010	Rare Earth Supply Technology and Resources Transformation Act of 2010	Rare Earths and Critical Materials Revitalization Actof 2010
Short Title	House RESTART Act	Senate RESTART Act	
Overall status	Early stage; first step in the legislative process	Senate subcommittee hearing held on September 30, 2010	Passed by House on September 29, 2010
Detailed status	**H.R. 4866** introduced on March 17, 2010, by **Rep. Mike Coffman** (R-CO), and referred to the following committees: House Armed Services Committee, House Ways and Means Committee, , House Financial Services Committee.	**S. 3521** introduced on June 22, 2010, by **Sen. Lisa Murkowski** (R-AK), and referred to the following committee: Senate Energy and Natural Resources Committee, Subcommittee on Energy. Hearing held by subcommittee on September 30, 2010.	**H.R. 6160** introduced on September 22, 2010 by **Rep. Kathleen A. Dahlkemper**(D-PA), and passed by House on September 29, 2010, and referred to Senate. Pending in the Senate Energy and Natural Resources Committee.
Goal	To establish a competitive domestic rare earth minerals production industry; a domestic rare earth processing, refining, purification and metals production industry; a domestic rare earth metals alloying industry; and a domestic rare earth based magnet production industry and supply chain in the United States.	To provide for the re-establishment of a domestic rare earth materials production and supply industry in the United States and for other purposes.	To develop a rare earth materials program, to amend the National Materials and Minerals Policy, Research and Development Act of 1980 and for other purposes.
Assessment of global market situation	Urgent need to identify global market situation regarding rare earths and the strategic values placed on them by other nations.	Urgent need to identify global market situation regarding rare earths and the strategic values placed on them by other nations.	
Proposed new government body/program	Interagency Working Group for purposes of re-establishing a competitive domestic rare earth supply chain; consisting of "Executive Agents" who are Assistant Secretary-level officials appointed by Secretaries of Commerce,	Rare Earth Policy Task Force to expedite permitting and projects that will increase exploration for, and development of, domestic rare earths. Task force established within Dept of Interior and Secretary of Interior reports	Under Section 101, the Act establishes in DOE a program of research, development, demonstration and commercial application to assure the long-term, secure and sustainable supply of rare earth materials sufficient to satisfy the national security, economic well-being

(Continued)

Legislation	House	Senate	House
	Defense, Energy, Interior and State. USTR and OSTP shall appoint representation to the WG.	to the President. Task Force composed of Secretaries Defense, Commerce, State, OMB, CEQ and other members as Secretary of Interior sees appropriate. Task Force to monitor and(or designee) from Interior, Energy, Agriculture, assist Fed agencies in expediting review and approval of permits or other actions, as necessary; assist Fed agencies in reviewing laws and policies that discourage investment in, exploration of and development of domestic rare earths. Annual report to the President and various House and Senate Committees.	and industrial production needs of the U.S. To the maximum extent possible, the Secretary shall support new or significantly improved processes and technologies as compared to those currently in use in the rare earth materials industry. TheSecretary shall encourage multidisciplinary collaborations among program participants and extensive opportunities forstudents at institutions of higher education, includinginstitutions listed under section 371 (a) of the Higher Education Act of 1965. The Secretary may collaborate on activities of mutual interest with the relevant agencies of foreign countries.
Specified Plan for implementation			Within 180 days after the enactment date of this Act, the Secretary shall prepare and submit to the appropriate Congressional committees a plan to carry out the program, including the criteria to be used to evaluate applications for loan guarantees under section 1706 of the Energy Policy Act of 2005. In preparing each plan the Secretary needs to consult with appropriate representatives of industry, institutions of higher education, DOE national labs, professional and technical societies and other entities, as determined by the Secretary.
Program assessment			After the program has been in operation for 4 years, the Secretary shall offer to enter into a contract with the National Academy of Sciences underwhich the National Academy shall conduct an assessment of the program. The assessment will include the recommendation from NASwhether the program should be continued with improvements or terminated with lessons learned. The assessment will be made available to Congress and the public upon completion.
Establish baseline for supply chain vulnerability	Secretaries of Commerce, Defense, Energy, Interior and State will undertake an assessment, coordinating with USTR and OSTP.	Secretaries of Interior and Energy will jointly, in consultation with Secretaries of Defense, Commerce, State and USTR, undertake an assessment, determining which rare earth elements are critical to clean energy techs and national and economic security of the U.S.	

(Continued)

Legislation	House	Senate	House
National Stockpile	Secretary of Defense shall commence procurement of rare earth materials designed as "critical" within one-year after enactment of this Act. Annual report and update on addition and subtraction.	Secretaries of Interior and Energy will jointly, in consultation with Secretaries of Defense, Commerce, State and USTR conduct an assessment as to whether rare earth elements determined critical need to be stockpiled or not and recommend criteria usedin determining the commencement and termination of stockpiling.	
Fair Market Conditions	USTR to initiate a review of international trade practices and initiate an action before the WTO, or issue a report to Congress after the review if action before WTO deemed not necessary.		
Loan Guarantees	Commerce, Interior and State to issue a report to industry describing government loan guarantees to reestablish a domestic rare earth supply chain. Secretary of Defense to issue guidance related to obtaining LGs under 50 U.S.C. 98 and any other available mechanisms. Secretary of Energy to issue guidance under the American Recovery and Reinvestment Act of 2009, Energy Efficiency and Renewable Energy sponsored programs and any other available mechanisms for obtaining LGs.	Secretary of Energy to issue report to the industry describing available mechanisms for LGs for establishing a domestic rare earth supply chain. Secretary of Energy to issue guidance for the rare earth industry on obtaining LGs under title XVII of the Energy Policy Act of 2005 and the American Recovery and Reinvestment Act of 2009.	Within 180 days after the enactment date of this Act,the Secretary shall prepare and submit to the appropriate Congressional committees a plan to carry out the program, including the criteria to be used to evaluate applications for loan guarantees under section 1706 of the Energy Policy Act of 2005. Title XVII of the Energy Policy Act of 2005 is amended by adding at the end the new section "Sec. 1706. Temporary Program for Rare Earth Materials Revitalization." The Secretary is authorized, to the extent provided in a subsequent appropriations act, to make guarantees for the commercial application of new or significantly improved technologies for projects that
			process rare earth materials, manufacture technologies that apply rare earth materials and other projects as determined by the Secretary. The authority to enter into guarantees under this section expires on September 30, 2015.
Defense Production Act	A prioritization of DPA projects to re-introduce a domestic rare earth supply chain. Secretary of Defense to issue a report describing past, current and future DPA projects for this purpose.	Secretary of Defense to submit to Congress a report describing past, current and future projects under the DPA of 1950 to support the domestic rare earth supply chain. If no such projects, need justifications for lack of such projects.	

(Continued)

Legislation	House	Senate	House
R&D	Commerce, Defense, Energy and Interior will use base budget funding to fund academic institutions, government labs, corporate R&D, non-profit R&D and industry associations.		The proposed DOE program shall support activities to: (1) Better characterize and quantify virgin stocks of rare earth materials using theoretical geochemical research; (2) explore, discover and recover rare earth materials using advanced science and technology; (3) improve methods for the extraction, processing, use,recovery and recycling of rare earth materials; (4) improve the understandingof the performance, processing and adaptabilityin engineering designs of rare earth materials; (5) identify and test alternative materials that can be substituted for rare earth materials in particular applications; (6) engineer and test applications that use recycled and alternative rare earth materials, and minimize rare earth materials content; (7) collect and disseminate info on rare earth materials; (8) facilitate info sharing and collaboration among program participants and stakeholders; (9) collaborate with foreign countries on R&D activities of mutual interest.
Innovation, Training and Workforce Development		Energy, Interior, Commerce and Defense should each provide funds to academic institutions, government labs, corporate R&D, non-profit R&D and industry associations in support of	
		innovation, training and workforce development in the domestic rare earth supply chain. The Depts. should give priority to academic institutions, government labs,corporations, non-profit entities and industry associations that will utilize domestically produced rare earths and associated materials.	
Amendments to the National Materials And Minerals Policy, Research and Development Act of 1980			Various amendments to the 1980 Act to ensure consistency with the newly inserted content of the Rare Earths and Critical Materials Revitalization Actof 2010.
Federal agency roles	Commerce, Defense, Energy, Interior, State, USTR, OSTP Major role: Commerce, Defense, Energy, Interior and State	Interior, Energy, Agriculture, Defense, Commerce, State, OMB, CEQ, USTR Major role: Interior, Energy, Defense	Energy

Appendix D. TREM Conference 2010 Address – Assistant Secretary David Sandalow

REMARKS PREPARED FOR DELIVERY

TECHNOLOGY AND RARE EARTH METALS CONFERENCE 2010

KEYNOTE ADDRESS

DAVID SANDALOW
ASSISTANT SECRETARY
FOR POLICY & INTERNATIONAL AFFAIRS
U.S. DEPARTMENT OF ENERGY

WASHINGTON, D.C.
MARCH 17, 2010

[Acknowledgements.]

1. INTRODUCTION

Thank you for the invitation to speak at this important conference.

At energy conferences today, no topic is hotter than shale gas. The story is striking: recoverable reserves of shale gas have increased six-fold in the past few years, thanks to new drilling technologies. This increase has been transformational, with U.S. natural gas imports now predicted to drop steadily in the next decade and beyond, whereas just a few years ago imports were projected to climb for the foreseeable future. Large shale gas reserves are believed to exist in other places around the world, including China and Eastern Europe, with potentially dramatic implications for patterns of energy use in those regions as well.

What does this have to do with rare earth metals?

Before answering that question, let me tell you about a much less well-known resource: rhenium.

Rhenium is the chemical element with atomic number 75. In part due to its very high melting point, rhenium is a critical component in nickel-based "superalloys," which are capable of functioning under very high stress. These superalloys are used in the jet engines of military aircraft and some of the world's most energy-efficient gas turbines.
However, rhenium is very rare. It is a byproduct from copper ores, but on average 120 tons of copper are needed to produce 30 grams of rhenium. In recent years, as rhenium's use in turbine blades and other applications has grown, its price has increased sharply.

So metallurgists at General Electric have worked to develop alternative superalloys that use less rhenium. GE has also partnered with the U.S. Navy and leading airlines to recycle turbine blades that contain rhenium. These steps have helped mitigate the risk of future rhenium supply disruptions or price spikes.

What do these two stories tell us?

That supply constraints aren't static. Strategies for addressing shortages of strategic resources are available, if we act wisely. We can invest in additional sources of supply. We can develop substitutes. We can re-use materials and find ways to use them more efficiently. We can consider use of stockpiles and strategic reserves. Not every one of these strategies will work every time. But taken together, they offer a set of approaches we should pursue as appropriate whenever potential shortages of natural resources loom on the horizon.

2. RARE EARTH METALS

That brings us to rare earth metals. As participants at this conference know, "rare earths" are typically defined to include the 15 elements in the periodic table with atomic numbers 57 through 71, along with several other elements. They include elements unfamiliar to most people, such as lanthanum, cerium, neodymium and europium.

Ironically, rare earth elements are not actually rare. Their misleading name probably comes from early mining of these elements in Europe, which focused on extraction from minerals that were quite uncommon. In fact, most rare earth elements are widely distributed in the Earth's crust. Indeed the abundance of rare earths in the Earth's crust is higher than that of some major industrial metals. Even the least abundant rare earths are found in greater quantities than, for example, bismuth and cadmium.

Rare earth elements have many desirable properties, including the ability to form unusually strong, lightweight magnetic materials when alloyed with other metals. They also have distinctive and valuable optical properties including fluorescence and emission of coherent light – important for lasers. Many of these properties result from the presence of an unfilled inner electron shell in their atomic structure.
These properties and others have made rare earth metals especially valuable in a number of applications, including for clean energy technologies. Lanthanum (atomic number 57) is used in batteries. Neodymium (atomic number 60) is used in magnets for electric motors. Europium (atomic number 63) is used in colored phosphors and lasers. Rare earth metals are also used in manufacturing energy-efficient windows and in capacitors, sensors and scintillators used in electricity transmission.

Although rare earth metals are found in many places on Earth, including the United States, Canada, Australia, Brazil and South Africa, they are difficult to extract in profitable quantities without substantial time and cost. This has led to geographically concentrated production.

Geographic concentration in the production of rare earth metals is not new. Prior to 1948, most of the world's rare earth production came from placer sand deposits in Brazil and India. In the following decades, production shifted to monazite deposits in South Africa and elsewhere. In the 1970s and 1980s, global production was dominated by the output from the Mountain Pass mine in California. Today, although production of rare earths still occurs in a number of countries, more than 95 percent of global supply is produced in China.

It goes without saying that diversified sources of supply are important for any strategic material. So too are substitutes and strategies for re-use and recycling. If rare earth metals are going to play an increasing role in our economy, we need to pursue those strategies. And there's every reason to believe that rare earth metals could play an increasing role in the global economy as the world transitions to clean energy.

3. CLEAN ENERGY ECONOMY

This transition is already well underway. The world is on the cusp of a clean energy revolution. Here in the United States, the Obama Administration is making historic investments in clean energy.

The American Recovery and Reinvestment Act was the largest one-time investment in clean energy in our nation's history – more than $80 billion. At the Department of Energy, we're investing our $37 billion in Recovery funds in electric vehicles; batteries and advanced energy storage; a smarter and more reliable electric grid; and wind and solar technologies, among many other areas. Through this investment, we'll at least double our renewable energy generation and manufacturing capacities by 2012. We'll also deploy hundreds of thousands of electric vehicles and charging infrastructure to power them; weatherize at least half a million homes; and expand our grid.

Other countries are also seizing this opportunity. Indeed, the market for clean energy technologies is growing rapidly all over the world.

Today, the Chinese government is launching programs to deploy electric cars in 13 major cities. It's connecting urban centers with high-speed rail. It's building huge wind farms, ultra-supercritical advanced coal plants and ultra-high-voltage long-distance transmission lines with low line loss.

India has launched an ambitious National Solar Mission, with the goal of reaching 20 gigawatts of installed solar capacity by 2020.

In Europe, strong public policies are driving sustained investments in clean energy. Denmark is the world's leading producer of wind turbines, earning more than $4 billion each year in that industry. Germany and Spain are the world's top installers of solar photovoltaic panels, accounting for nearly three-quarters of a global market worth $37 billion last year.
Around the world, investments in clean energy technologies are growing, helping create jobs, promote economic growth and fight climate change. These technologies will be a key part of the transition to a clean energy future.

However today, many of these technologies rely on the special properties of rare-earth metals. There's no reason to panic, but there's every reason to be smart and serious as we plan for growing global demand for products that contain rare earth metals and other strategic materials.

4. STRATEGIES TO MANAGE POTENTIAL RISKS

For the clean energy economy to reach its full potential, we must work together to ensure stable supplies of the materials required. That means working together to diversify global supply chains, as well as investing in manufacturing and processing. It means research and development into substitutes. It means finding ways to recycle and re-use scarce materials. U.S. talent and innovative capacity in materials science can be harnessed to create the next generation of rare earth applications and competing technologies.

To proactively address the availability of rare earths and other strategic materials required for the clean energy economy, we must take a three-part approach:

The first strategy is to **globalize supply chains for strategic materials**. To paraphrase what Churchill once said about oil: Security rests above all in diversity of supply. To manage supply risk, we need multiple, distributed sources of strategic materials in the years ahead. This means taking steps to encourage extraction, refining and manufacturing here in the United States, as well as encouraging our trading partners to expedite the environmentally-sound creation of alternative supplies.

The United States will explore investments at all stages of the supply chain, from environmentally-sound material extraction, to purification and processing, to the manufacture of chemicals and components, and finally to end uses. These investments will help the United States strengthen our manufacturing base over the long term. And we will examine issues related to strategic stockpiling of critical materials, especially those with dual-use potential, i.e., in both civilian and military applications.

Second, we must **develop substitutes**. Doing so will improve our flexibility as we address the materials demands of the clean energy economy. This means investing in RD&D to develop transformational magnet, battery terminal and other technologies that reduce our dependence on rare earths. DOE's Vehicle Technologies Program, the ARPA-E program, and our national labs are all currently conducting research along these tracks. The ARPA-E program recently funded a consortium of universities, laboratories and private companies to conduct research into high-energy permanent magnets that will use domestically-available materials while more than doubling magnetic energy density over current state-of-the-art technologies.

Third, we must promote **recycling, re-use and more efficient use** of strategic materials, to get more economic value out of each ton of ore extracted and refined. As in the case of GE's turbine blade recycling program, re-use can help mitigate potential supply constraints. Widespread recycling and re-use could significantly lower world demand for rare earths and other strategic materials. And, importantly, recycling and re-use could also reduce the lifecycle environmental footprint of these materials.

As a society, we have dealt with these types of issues before, mainly through smart policy and R&D investments that reinforced efficient market mechanisms. We can and will do so again.

To help address these concerns, I am today announcing that the Department of Energy will develop its first-ever strategic plan for addressing the role of rare earth and other strategic materials in clean energy technologies. The plan will apply the approaches described above and draw on the strengths of the Department in technology innovation. We will build on work on these topics already underway, including in DOE's national labs, and work closely with colleagues from other agencies throughout the U.S. government. We will solicit broad public input, including from the stakeholders and experts here in this room.

5. CONCLUSION

So in conclusion, as I said earlier: there's no reason to panic, but every reason to be smart and serious as we plan for growing global demand for products that contain rare earth metals.

The United States intends to be a world leader in clean energy technologies. Toward that end, we are shaping the policies and approaches to help prevent disruptions in supply of the materials needed for those technologies. This will involve careful and collaborative policy development. The United States will develop and implement systematic approaches to building a stable, geographically diverse supply chain; encourage technical innovations to identify substitutes as well as minimize the requirements for these key materials; and encourage recycling and re-use wherever possible. We will rely on the creative genius and entrepreneurial ingenuity of the business community to meet an emerging market demand in a competitive fashion. Working together, we can meet these challenges.

Appendix E. U.S.–Japan Roundtable on Rare Earth Elements Research and Development for Clean Energy Technologies Agenda

Agenda

U.S. – Japan Roundtable on Rare Earth Elements Research and Development for Clean Energy Technologies

Thursday-Friday, November 18-19, 2010
Building 453, Black Diamond Conference Room 1012

U.S. Department of Energy
Lawrence Livermore National Laboratory

THURSDAY, November 18, 2010

7:30 **Badging Westgate Badge Office**
Met by Evelyn Laurant

8:30 – 9:00 **Welcome and Introduction to the Roundtable**
Al Ramponi, Lawrence Livermore National Laboratory
Kay Thompson and Diana Bauer, Department of Energy

9:00 – 9:30 **A Brief Overview of the Rare Earths Crisis**
Karl Gschneidner, Ames Laboratory
Discussant: Roderick Eggert, Colorado School of Mines

9:30 – 10:45 **Geological Availability of Rare Earth Elements**
Session Chair: Bill Bourcier, Lawrence Livermore National Laboratory
Keith Long, U.S. Geological Survey; *New Sources of Primary REE Production: When and at What Cost*
Tetsuichi Takagi, National Institute of Advanced Industrial Science and Technology, Japan
Bradley Van Gosen, U.S. Geological Survey; *The Principal Rare Earth Elements Deposits of the United States*

10:45 – 11:00 **Break**

11:00 – 12:30 **Recovery, Extraction and Separation of REE from Mineral Ores, Part I**
Session Chair: Alex King, Ames Laboratory
Brock O'Kelley, Molycorp Minerals, LLC
Patrick Taylor and Corby Anderson, Colorado School of Mines; *Mineral Processing, Extraction and Refining of Rare Earth Minerals*
Junji Shibata, Kansai University; *Separation and Purification Technology for REE*

12:30 – 1:30 Lunch
Group to walk over to Central Cafeteria

1:30 – 2:45 Recovery, Extraction and Separation of REE from Mineral Ores, Part II
Session Chair: Christian Mailhiot, Lawrence Livermore National Laboratory
Eric Peterson, Idaho National Laboratory (INL); *INL Separations and Lanthanide Chemistry Activities*
John Hryn, Argonne National Laboratory; *Solvent Extraction for Separation of REE*
Patrick Huang, Lawrence Livermore National Laboratory; *Computational Approaches to f-electron Chemistry*
Discussant: Suresh Baskaran, Pacific Northwest National Laboratory

2:45 – 3:00 Break

3:00 – 4:30 Improved Manufacturing and Use of REE
Session Chair: Mark Rigali, Sandia National Laboratories
John Hryn, Argonne National Laboratory; *Improved Manufacturing and Use of REE*
Karl Gschneidner, Ames Laboratory; *A New, Energy Efficient Green Process for Preparing Commercial Grade Rare Earth Metals Including Nd and La*
Discussant: Mike McElfresh, National Security Technologies, LLC
Discussant: Suresh Baskaran, Pacific Northwest National Laboratory

4:30 Adjourn for the day

5:30 Dinner Terra Mia restaurant, 4040 East Avenue, Livermore
By registration

FRIDAY, November 19, 2010

8:30 – 9:00 Overview of New Energy and Industrial Technology Development Organization Effort
Toru Nakayama, New Energy and Industrial Technology Development Organization

9:00 – 9:30 Advanced Energy Research Projects Agency Perspective on the Rare Earth Materials Issue
Mark Johnson, Advanced Energy Research Projects Agency

9:30 – 10:30 Alternatives and Substitutes for REE Technologies
Co-Chairs: Roderick Eggert, Colorado School of Mines and Koki Hanzawa, New Energy and Industrial Technology Development Organization
Tomoyuki Ogawa, Tohoku University; *Development of High Performance Magnetic Materials with Rare Earth Element (REE) Less/Free*
Bill McCallum, Ames Laboratory; *Magnets*

Kimihiro Ozaki, National Institute of Advanced Industrial Science and Technology, Japan; *Research and Development of Dy-free Rare Earth Magnets as Alternatives to Nd-Fe-B Magnets*

10:30 – 10:45 Break – Group Photo

10:45 – 12:30 Alternatives and Substitutes for REE Technologies, cont'd
Atsushi Muramatsu, Tohoku University; *REE Engineering Overview and Cerium: its Alternative and/or Curtailment Technologies to Conventional Method*
John Hryn, Argonne National Laboratory; *Alternatives and Substitutes for REE*
Momoji Kubo, Tohoku University; *Computational Chemistry as a Powerful Tool for the Design of the REE Alternative and/or Curtailment Technologies and the Integrated Application of the Experiments and Simulations to Cerium for the Mechanical Polishing*
Yutaka Tai, National Institute of Advanced Industrial Science and Technology, Japan; *Ceria in Automotive Catalysts*

12:30 – 1:30 Lunch
Group to walk to Central Cafeteria

1:30 – 3:00 Summarize Findings and Recommendations of Roundtable

3:00 Closing Remarks and Adjourn

APPENDIX F. TRANS-ATLANTIC WORKSHOP ON RARE EARTH ELEMENTS AND OTHER CRITICAL MATERIALS FOR A CLEAN ENERGY FUTURE

US-EU ENERGY COUNCIL

Trans-Atlantic Workshop on Rare Earth Elements and Other Critical Materials for a Clean Energy Future

Hosted by the MIT Energy Initiative
Massachusetts Institute of Technology
400 Main Street, Building E19-307, Cambridge, Massachusetts
December 3, 2010

Workshop Background

Rare earth elements and other critical raw materials are essential to our industrial production, particularly for clean energy options like wind turbines, solar cells, electric vehicles, and energy-efficient lighting. Wind turbines are the most rapidly growing source of electricity generation in both Europe and the United States. Solar photovoltaic cells are steadily declining in cost, and their widespread, cost-effective use on power grids is anticipated within the coming decade. Electric vehicles, meanwhile, offer a means to move away from imported oil for transport towards a mix of coals, gas, wind, solar and nuclear energy with much lower net carbon dioxide emissions. Compact fluorescent and LED lighting offer an avenue to greatly reduce electricity consumption.

Yet these vital clean energy options will use a large share of available rare earths and other less common materials. Production of some of these materials is concentrated in a very small portion of the globe, so that supplies may become tight and costs prohibitive as markets grow. To cope with this dilemma, there are a few main effective strategies. First, we can try to find new or enhanced recycling technologies to increase available supplies. Second, we can try to find substitute materials or alternate device designs that perform as well or nearly as well at comparable cost. Third, we can look for technology and process design changes to limit the amounts of scarce materials that are required.

Objectives

The workshop will gather experts from the US and the EU to exchange views on emerging challenges emerging form scarce availability of rare earths and other critical elements.

What are the most important materials for continued expansion of clean energy markets? Which of these materials are likely to experience supply constraints over the next two decades? Which are the priorities for research, particularly for the substitution of the use of critical elements? To what extent can we alleviate tight supplies through enhanced exploration and development? How might wind turbines and electric vehicles be redesigned so they do not rely on scarce materials? What kinds of advanced materials can substitute for the materials now in use? What new technology pathways should we follow to find the substitutes we need? What are the opportunities for Trans-Atlantic cooperation to accelerate our progress along those pathways and leverage the substantial resources that are being devoted to them?

These are the types of questions that will be addressed by the participating top scientists, engineers, officials, and utility executives with expertise in advanced materials, materials recycling, electric vehicle drives, wind turbine design, and photovoltaic power systems to discuss these issues, define the most promising technology options, and assess the long term research and development opportunities. Interactions should help examining the practical potential to achieve clean energy functionality with prudent use of critical materials, through substitutes, alternate technologies and more efficient use.

Workshop Venue and Organization

The workshop will be hosted by the US Department of Energy (DOE) and the Massachusetts Institute of Technology (MIT) in Cambridge, Massachusetts on December 3, 2010 and is organized in cooperation with the European Commission Directorate General for Research and Innovation, Unit "Materials", under the auspices of the EU-US Energy Council, its Working Group on Energy Technology, and its expert subgroup on advanced materials. This initiative also contributes to the work developed within the Innovation Action Partnership under the EU US Transatlantic Economic Council.

The workshop will be limited in size, and all participants are expected to engage actively in discussion. Officials of the European Commission and DOE, invited selected high level scientists and experts to participate.

A Keynotes: Setting the Scene - Critical Materials for a Clean Energy Future

(8:30 - 10:30 am)

Keynote addresses will be invited from leading energy policy makers, scientists and executives to highlight findings of the EU Report on Critical Materials, the DOE Critical Material Strategy, and the MIT report on Critical Elements in Energy Technology, all recently released or about to be released. A Japanese keynote speaker is invited.

Chairs/Animateurs: **Jeff Skeer**, DOE Office of Policy and International Affairs and **Renzo Tomellini**, EC Directorate General for Research and Innovation

Rapporteurs: **Tom Lograsso,** The Ames Laboratory and **Nick Morley**, Oakdene Hollins.

Each Speaker will have 15 minutes

Diana Bauer, Office of Policy and International Affairs, U.S. Department of Energy, *Highlights of the DOE Critical Materials Strategy*

Antje Wittenberg, Directorate General for Enterprise and Industry, *The EU Raw Materials Initiative and the Report of the Ad-hoc Group (tbc)*

Tom Lograsso, The Ames Laboratory (Iowa), *Future Directions in Rare Earth Research: Critical Materials for 21st Century Industry*

Derk Bol, Materials Innovation Institute M2i (Netherlands) M2i, *Material Scarcity Report and Industrial Perspectives*

Bob Jaffe, Massachusetts Institute of Technology, *Insights from the Energy Critical Elements Policy Study by the American Physical Society and Material Research Society*

Evangelos Tzimas, EC Joint Research Centre, Institute for Energy, *Preliminary Findings on the Role of Rare Metals as Supply Chain Bottlenecks for Priority Energy Technologies (tbc)*

Kazuhiro Hono, Magnetic Materials Center Managing Director, NIMS, *Research Trends on Rare Earth Materials in Japan*

Edward Jones, Lawrence Livermore National Laboratory, *Outcomes of U.S.-Japan Roundtable on Rare Earth Elements R&D for Clean Energy Technologies (18-19 November 2010)*

QUESTIONS AND ANSWERS TIME AND DISCUSSION

Coffee Break (10:30 – 11:00 am)

B Strategies and Research for Finding Critical Material Substitutes

(11:00 am - 12:30 pm)

This session will highlight ways to substitute for critical materials that are used in clean energy devices like EV motors, LED lighting and solar cells. How can science find materials that substitute for those used in clean energy devices today and perform just as well? How can R&D efforts help us design devices that perform the same function with relatively inexpensive and easily available materials?

Chairs/Animateur: Linda Horton, DOE Office of Basic Energy Sciences

Rapporteurs: **Tom Lograsso,** The Ames Laboratory and **Nick Morley**, Oakdene Hollins

Each Speaker will have 10 minutes

George Hadjipanayis, Chairman, Department of Physics and Astronomy, University of Delaware, *Moving Beyond Neodymium-Iron Permanent Magnets for EV Motors*

Bertrand Fillon, Commissariat à l'Energie Atomique et aux Energies Alternatives, *Challenges for the Future Energy Generation, Distribution and Use*

John Hsu and/or **Tim Burress,** Oak Ridge National Laboratory, *Flux Coupling Machines and Switched Reluctance Motors to Replace Permanent Magnets in Electric Vehicles*

Spomenka Kobe, Jozef Stefan Institut, *Rare Earth Magnets in Europe*

Madhav Manjrekar, Green Energy and Power Systems, Siemens Corporate Research, *Research Priorities for Critical Material Substitutes from a European Corporate Perspective*

Anne de Guibert, SAFT, *Critical Materials and Alternatives for Storage Batteries*

QUESTIONS AND ANSWERS TIME AND DISCUSSION (30 minutes)

Lunch (12:30 – 1:30 pm)

Hosted by the Delegation of the European Union to the United States of America

C Strategies and Research for Using Critical Materials More Effectively

(1:30 - 3:00 pm)

This session will highlight strategies for reducing critical material needs over multiple device lifecycles. These include innovations to *reduce waste* in manufacturing, *enhance recycling* of critical materials used, and *design molecular structure* so that less critical material is required.

Chairs/Animateur: **Pilar Aguar**, EC Directorate General for Research and Innovation

Rapporteurs: **Tom Lograsso,** The Ames Laboratory and **Nick Morley**, Oakdene Hollins

Each Speaker will have 10 minutes

Iver Anderson, Division of Materials Sciences and Engineering, The Ames Laboratory, *Current and Future Direction in Processing Rare Earth Alloys for Clean Energy Applications*

Michael Heine, SGL Group - The Carbon Company, *Carbon Fibers in Lightweight Systems for Wind Energy and Automotive Applications: Availability and Challenges for the Future*

Steve Duclos, Chief Scientist, GE Global Research, *Research Priorities for More Efficient Use of Critical Materials from a U.S. Corporate Perspective*

Christian Hagelüken, UMICORE, *"Opportunities and Limits to Recycling of Critical Materials for Clean Energies*

Peter Dent, Electron Energy Corporation, *Strategies for More Effective Critical Materials Use*

Daniel Beat Müller, Norwegian University of Science and Technology, *Material Flow Analysis*

QUESTIONS AND ANSWERS TIME AND DISCUSSION (30 minutes)

Refreshment Break (3:00 – 3:30 pm)

D Opportunities for EU-US Cooperation on Critical Energy Materials

(3:30 - 4:30 pm)

This session will identify synergies between U.S. and EU efforts to find substitutes for critical materials and reduce the needs for such materials in clean energy technologies. These synergies will form the core of a Trans-Atlantic Strategic Vision for Critical Materials Research.

Chairs/Animateurs: **Jeff Skeer**, DOE Office of Policy and International Affairs and **Renzo Tomellini**, EC Directorate General for Research and Innovation

Rapporteurs: **Tom Lograsso,** The Ames Laboratory and **Nick Morley**, Oakdene Hollins

E. Wrap Up Session

(4:30 – 5:00 pm)

This session will identify next steps and actions in pursuing a collaborative research agenda.

Chairs/Animateurs: **Jeff Skeer**, DOE Office of Policy and International Affairs and **Renzo Tomellini**, EC Directorate General for Research and Innovation

Rapporteurs: **Tom Lograsso,** The Ames Laboratory and **Nick Morley**, Oakdene Hollins

Close (5:00 pm)

APPENDIX G: ARPA-E WORKSHOP AGENDA

BACKGROUND

The workshop will consist of five plenary talks and four breakout sessions. The breakout sessions will be run concurrently in groups of two, with focuses on:

Supply / Material processing

Magnets and magnetic systems / motors and generators

Phosphors and general illumination

Catalysts and separators

Experts from across science and engineering will be brought together to identify possible new approaches and pathways to addressing technical challenges in critical materials across these fields.

SUPPLY / MATERIALS PROCESSING

BACKGROUND

Existing availability of rare-earth materials is limited by the processing of these materials into usable form (elemental metal, oxide etc) from known deposits. Existing process technologies are based on acid-alkaline chemical processes followed by solvent extraction to yield elemental species separation. Commercial scale acid-alkaline processes use large quantities of reagents. As a result, environmental and cost considerations for processing rare-earth materials have been prohibitive, particularly in developed nations. Environmental impacts are significant with more tangible costs in developed nations. Pathways to new rare-earth materials processes, which have minimized environmental impacts, are of high interest for discussion and examination in this workshop.

The rare earths are rather challenging to separate chemically as the latent heats of oxide formation are nearly the same between rare-earth elements. Further, solubilities of rare-earth ions in solution and separation coefficients are very similar for most solvents. Rare earths elements may often be separated into light (lanthanum, cerium, neodymium, praseodymium) and heavy (terbium, europium lutetium, gadolinium) elements based on characteristic ores. Light rare-earth element deposits are by far more prevalent. In this workshop and break out session, we are interested in completely new processes and approaches (physical, chemical or biological) to separation and processing, particularly for heavy rare-earth elements of high importance to energy applications.

With the increased demand for neodymium and dysprosium for permanent magnets, new high-selectivity separation technologies are needed. Neodymium has low separation selectivity relative to praseodymium, and dysprosium occurs in very low concentration in most rare-earth ores. New separation technologies might be based on chemical (species specific ionic liquids and solvents), physical (species specific porous membranes) or biological (elementally specific geobacters) means as possible mechanism for processing. New technology pathways to cost effective separation and concentration of neodymium and dysprosium are of high interest for examination in this workshop.

QUESTIONS:

Are there completely new process pathways which may be used, and subsequently scaled, for the processing of rare earths?

Can we leverage the knowledge base from the actinide materials processes realm to the lanthanide / rare-earth materials processes? For example, highly selective processes for the isolation of Americium and for the separation of Thorium from water have been developed. Can the physical and chemical methods used in these heavy-element separation technologies be applied to rare-earths?

Geobacters have been studied over the past few years as microorganisms with metabolisms which chemically interact with compounds and minerals. Can geobacters with strong affinity for specific

MAGNETICS

BACKGROUND

Rare earth metals of Neodymium and Dysprosium are of high importance for highest energy product permanent magnets, in the form of $Nd_2Fe_{14}B$ material. Such magnets are of increased use in electric motor and generator applications of the energy sector as the coupling means between torque and electricity. Dysprosium is used in higher Curie temperature permanent magnets and as a result is of increased importance for electric motors, particularly in the transportation sector. In this workshop, magnetics will be addressed from two directions: identification of potential material alternatives at the component level and of potential magnet replacements at the system level. Technologies for either possible technical approach (component or system) are of high interest and possible pathways will be investigated in this workshop.

Nanoscale magnetic material mixtures have been shown to have potential for high-energy product in permanent magnets, with reduced rare earth material content, through spring-exchange coupling of hard and soft magnetic phases in close proximity (< 50nm). ARPA-E has already initiated two projects in this approach: University of Delaware project is focused on aligned nanoscale mixtures of hard and soft magnetic materials; and General Electric project will demonstrate consolidated hard and soft phases of core/shell nanoparticles

In this workshop and breakout session, we are looking for even newer / alternative approaches to reduced rare-earth content permanent magnets in energy systems.

QUESTIONS:

Are there other non-rare earth content magnets which may exhibit high coercivity, particularly as a potential hard-phase in nanoscale composites? For instance PtCo forms very high energy density permanent magnets with no rare-earth content, although the cost is prohibitive. What new magnets might exhibit high coercivity with low rare-earth content?

Alternatively, what are the technical barriers to the use of superconducting magnets in some applications, such as high power wind turbines (>1 MW)?

Are permanent magnets necessary in high energy density electric motors for vehicles and what alternatives exist? For instance, with recent advances in power electronics, can induction or reluctance motorsefficiently couple torque and electric energy at acceptable cost and weight requirements? What technology advances are required to use non-permanent magnet motors in transportation or heavy industrial applications?

PHOSPHORS

BACKGROUND

Rare earth materials are used as both dopant or compositional quantities for phosphors. Terbium, Erbium, Yttrium and Europium are all used as red green and blue phosphors for white lighting. New phosphors are used to downshift UV or blue emission from LEDs to form color corrected white LEDs. Proposed strategies for dealing with reduced rare-earth availability include recycling programs where the phosphors from older tubes form the supply for new high-efficiency lighting.

In this workshop and breakout session, we are most interested in identifying over-the-horizon new technology opportunities for rare-earth free phosphors or alternative mechanisms for generation of high color rendering index white emission for high efficiency general illumination. Technologies which harnessing surface plasmon effects are of high interest.

QUESTIONS:

Are there high-efficiency optical transitions in non-rare-earth materials which can be used as an alternative particularly for new LED or OLED sources of white light? For instance, can size and shape of nanoparticle emitters be manipulated to control color in non-rare-earth phosphors as they are in aqueous environments for biological applications? Can new phosphors be developed which only use rare earths materials in dopant quantities (less than 0.1%)? For downshifted LED or OLED emitters, what technology gaps are there in matching emission wavelength to phosphor absorption to achieve high color rendering index white emission? What are the technology gaps for the emission intensity from multiple wavelength LEDs to be controlled, at a low enough cost, such that the resulting color remains consistent over time, even as individual emitters are subject to degradation?

CATALYSTS AND SEPARATORS

BACKGROUND

Rare earth oxides such as Ceria are used significantly in applications involving oxygen catalysis and transport through ceramic separators. Examples include: Ceria stabilization on Na-Y Zeolite catalysts are used for fluid catalytic cracking in most modern petroleum refining of gasoline for the transportation sector; Ceria (as well as Pt and Rh) form the basis for the 'three-way catalyst' in widespread use for post-combustion catalytic conversion for automobiles; and Ceria is used in yttria-stabilized zirconia (YSZ) separators used for high temperature diffusion electrolyte separators used in chemical processes and fuel cells.

In this workshop and breakout session, we are interested in exploring over-the-horizon new technical solutions which would provide alternatives or completely new pathways, to the use of rare-earth materials in catalysis and separator applications.

QUESTIONS:

Are there alternatives to the high use of rare-earths in fluid catalytic cracking? For instance mesoporous zeolites have been shown to increase both the efficiency and mass transfer rates for fluid catalytic cracking. What are the limits in controlling size and shape to dramatically increase unit process efficiency and throughput? What are the ultimate limits and by what quantitative amount could the process be improved while reducing rare-earth content? In high diffusivity oxygen separators, what other solutions might provide the combination of oxygen permeability and mechanical integrity at high temperatures, without rare-earth or critical materials? Finally, low or reduced dimensional materials such as graphene or nanotubes exhibit interesting catalytic properties, for instance on the edges and on the surfaces. Can these unique structures be developed into new catalyst technologies of significant importance for the energy sector?

BREAKOUT SESSIONS AND PARTICIPANT PREPARATION

Four breakout sessions of 90 minute lengths will be held through the workshop, in two periods of two concurrent sessions. Each participant is has been assigned a recommended breakout session to attend with an open option to attend either session during the second time period. Approximately 10 participants have been assigned to each session.

For the breakout session which a participant has been asked to attend, each participant is asked to prepare an up to 5 minute and 3 PowerPoint slide contribution, intended to spur discussion. Following two 15 minute stage-setting overview presentations, each participant will be asked for their contribution, interspersed with technical dialog between participants. All contributions will be accommodated, to the extent possible.

****** Any presentations created for the workshop should be publicly releasable ******

Private one-on-one sidebar meetings will be held with ARPA-E Program Directors both the night before, Sunday night, and at the conclusion of the day in 15 minute increments in the hotel restaurant area. The purpose of sidebar meetings is to share innovative technology concepts with Program Directors in a more private venue. In order to accommodate travel schedules at the end of the workshop, participants are encouraged to contact Angela Huffaker to schedule a time in advanced of the workshop, or sign up at the time of meeting.

AGENDA

**** Sunday Dec. 5 @ 6pm – Dinner in hotel with side-bar discussions for those arriving early. ****

8:00 AM	Registration / Coffee / Breakfast	
8:15 AM		
8:30 AM	Kick-Off	
8:45 AM	Overview of Day	
9:00 AM	US-Japan Rapporteur	
9:15 AM	US-EU Rapporteur	
9:30 AM	Current State of the Art for Supply	
9:45 AM	Current State of the Art for Applications Talk #1	
10:00 AM	Current State of the Art for Applications Talk #2	
10:15 AM	Break	
10:30 AM	SUPPLY Briefing #1	MAGNETS Briefing #1
10:45 AM	SUPPLY Briefing #2	MAGNETS Briefing #2
11:00 AM	SUPPLY Brainstorming / Dialog	MAGNETS Brainstorming / Dialog
11:15 AM		
11:30 AM		
11:45 AM		
12:00 PM		
12:15 PM	Break for Lunch	
12:30 PM	Keynote	
12:45 PM		
1:00 PM	Break	
1:15 PM	CATALYSTS Briefing #1	PHOSPHORS Briefing #1
1:30 PM	CATALYSTS Briefing #2	PHOSPHORS Briefing #2
1:45 PM	CATALYSTS Brainstorming / Dialog	PHOSPHORS Brainstorming / Dialog
2:00 PM		
2:15 PM		
2:30 PM		
2:45 PM		
3:00 PM	Break	
3:15 PM		
3:30 PM	Report Out	
3:45 PM		
4:00 PM		
4:15 PM		
4:30 PM	Conclusion of Day	
4:45 PM	Adjourn to "No Host" Happy Hour / Dinner (Hotel Bar) Sidebar Discussions with ARPA-E Program Director (On Request)	
5:00 PM		

CONTACT INFORMATION

MARK JOHNSON
ARPA-E Program Director
Mark.a.Johnson@hq.doe.gov
202-287-6084

ANGELA HUFFAKER
Contractor/Programmatic Support
Angela.Huffaker@hq.doe.gov
202-287-5998

REFERENCES

[1] NAS (National Academy of the Sciences). (2008). Minerals, Critical Minerals and the U.S. Economy.

[2] Arnold Magnetic Technologies. (2010) "Response *to Department of Energy Request for Information*." June 7.

[3] Brodd, R. J. (2005). *Factors Affecting U.S. Production Decisions: Why are there no Volume Lithium-Ion Battery Manufactureres in the United States*. Gaitherburg, MD: National Institute of Standards and Technology.

[4] Dent, P. C. (2009). "High Performance Magnet Materials: Risky Supply Chain" *Advanced Materials and Processes*, August. http://www.electronenergy. com/ media

[5] Deutsche Bank. (2009). *Electric Cars: Plugged In 2*. Frankfurt: Deutsche Bank Securities.

[6] DOE (U.S. Department of Energy). (2009). *Buildings Energy Data Book.* Washington, DC: DOE. http://buildingsdatabook.eren.doe.gov/Default.aspx.

[7] DOE (U.S. Department of Energy), EERE (Energy Efficiency and Renewable Energy). (2008). "Using Light Emitting Diodes: How LEDs Work." October 28. http://www1.eere.energy

[8] EERE (Office of Energy Efficiency and Renewable Energy) Solar Energy Technologies Program. (2010). "DOE $1/W Workshop Photovoltaic Industry Primer." Washington, DC: U.S. Department of Energy.

[9] Electron Energy Corporation. (2010). "Response to Department of Energy Request for Information." June 7.

[10] Gains, L. & Nelson, P. (2010). *Lithium-Ion Batteries: Examining Material Demand and Recycling Issues*. Argonne, IL: Argonne National Laboratory.

[11] Hamilton, T. (2009). "Lithium Battery Recycling Gets a Boost." *Technology Review*, August 29.

[12] Kingsnorth, D. (2010). "Rare Earths: Facing new Challenges in the new Decade." Slideshow presented at the 2010 Society for Mining, Metallurgy, and Exploration Annual Meeting, Phoenix. http://www.smenet.org/rareEarthsProject/SME_2010_Kingsnorth.pdf.

[13] Kopera, J. J. (2004). *Inside the Nickel Metal Hydride Battery.* Orion, MI: Cobasys. http://www.cobasys.com/pdf/tutorial/InsideNimhBattery/inside_nimh_battery_technology

[14] London, I. M. (2010). "The Delicate Supply Balance and Growing Demand for Rare Earths." Slideshow presented at Magnetics Economic Policy Photovoltaic Manufacturing Symposium, Washington, DC, July 29.
[15] Vehicle Technologies Program. (2009). "Batteries." U.S. Department of Energy, Office of Energy Efficiency and Renewable Energy. *Accessed July* 20, 2010. http://www1.eere.energy
[16] USGS (U.S. Geological Survey). (2009a). *2008 Minerals Yearbook: Lithium.* Reston, VA: USGS.
[17] USGS (U.S. Geological Survey). (2009b). *2008 Minerals Yearbook: Selenium and Tellurium.* Reston, VA: USGS.
[18] Arnold Magnetic Technologies. (2010). "Response to DOE Request for Information." June 5, 2010.
[19] Baylis, R. (2009). "*From Zhejiang to Zabuye: China's Mountainous Path to A Self-sufficient, Integrated, Lithium Industry*," Presented at the LSM Conference, Santiago, *January*.
[20] British Geological Survey. (2010). *Rare Earth Elements. Nottingham, UK: Natural Environment Research Council.* http://www.bgs.ac.uk/downloads/start.cfm?id=1638
[21] Chen, Z. (2010). "The development and policies of China rare earth industry." Presented to the 6th International Conference on Rare Earth Development and Application, August 2–6, Beijing. http://www.cs-re.org.cn/en/modules
[22] "China May Issue 12th Five-year Plan for Rare Earth Industry in September." (2010). Business China. July 26. http://en.21cbh.com/HTML/2010-7-26/China-Rare%20earth.html.
[23] Cobalt Development Institute. (2009). Cobalt Supply and Demand 2009. Cobalt Facts. http://www.thecdi.com/cdi/images
[24] Cobalt Development Institute. 2010. "Response to DOE Request for Information." June 7.
[25] Eggert, P. G. (2010). "Critical Minerals and Emerging Technologies." *Issues in Science and Technology*, Summer.
[26] General Electric, (2010). Submission to U.S. Department of Energy in Response to Request for Information on Rare Earth Elements and Other Technology Metals.
[27] Hatch, G. (2010). "Effect of China's REE Export Cuts." *The Gold Report.* http://www.theaureport.com/pub/na/6821.
[28] Humphreys, D. Forthcoming (2011). "Pricing and Trading in Metals and Minerals," *In SME Mining Engineering Handbook.* Littleton, CO: Society of Mining, Metallurgy, and Exploration.
[29] IMCOA. (2010). *Rare Earth Market Analysis*. "Submission by the company Molycorp as part of Molycorp's response to DOE Request for Information." May.
[30] Kingsnorth, D. (2010). "Rare Earths: Facing New Challenges in the New Decade." Presented at the annual meeting for the Society for Mining, Metallurgy, and Exploration, Phoenix, February 27– March 4.
[31] Kingsnorth, D. & Chegwidden, J. (2010). "*Rare Earths: Facing the Uncertainties of Supply.*" Presented at the 6th International Rare Earth Conference, Hong Kong, November.
[32] Koven, P. (2010). "China Imposes Export Quotas on Rare Earth Elements." *Financial Post*, July 12.

[33] Lynas Corp. (2010). What are their prices? Under What Are Rare Earths? http://www.lynascorp.com/page.asp?category_id=1&page_id=25
[34] Lynas Corp. 2010a Annual Report (2009). http://www.lynascorp.com/content/upload/files/Reports/Annual_Report_2009_778195.pdf.
[35] MBendi Information Services. (2010). "*World Nickel and Cobalt Mining - Mining and production*." http://www.mbendi.com/indy/ming/nkcb/p0010.htm.
[36] McMahon, Fred and Miguel Cervantes. (2010). *2009/2010: 2010 Mid-Year Update. Frazier Institute*. Van Couver, BC. http://www.fraserinstitute.org/uploadedFiles/fraser-ca/Content/researchnews/research/publications/miningsurvey-2010update.pdf.
[37] Molycorp. (2010). " Rare Earth Minerals:The Indispensable Resource for Clean Energy Technologies." Presented at the 2nd Annual Strategic Materials Conference, DMTC, Cleveland, OH. February 1-2, 2010.
[38] Roskill. (2009). "The Economics of Lithium," summary only. http://www.roskill.com/reports/minorand-light
[39] Roskill. (2010). "Rare Earths: An Evaluation of Current and Future Supply." Presentation by Judith Chegwidden to the 2010 Rare Earth Summit, Beijing, August 4-5.
[40] Simbol Mining. (2010). "Response to DOE Request for Information." June 7.
[41] Umicore USA. 2010. "Response to DOE Request for Information." June 6.
[42] USGS (U.S. Geological Survey). 1994–2010. *Mineral Commodity Summaries*: Rare earths. Reston, VA: USGS.
[43] USGS (U.S. Geological Survey). 1994–2010a. *Mineral Commodity Summaries*: Lithium. Reston, VA: USGS.
[44] USGS (U.S. Geological Survey). 1994–2010b. *Mineral Commodity Summaries*: Cobalt. Reston, VA: USGS.
[45] USGS (U.S. Geological Survey). 1994–2010c. *Mineral Commodity Summaries*: Gallium. Reston, VA: USGS.
[46] USGS (U.S. Geological Survey). 1994–2010d. *Mineral Commodity Summaries*: Indium. Reston, VA: USGS.
[47] USGS (U.S. Geological Survey). 1994–2010e. *Mineral Commodity Summaries*: Tellurium. Reston, VA: USGS.
[48] USGS (U.S. Geological Survey). 2004f. *USGS Minerals Yearbook 2004 Volume I: Cobalt chapter*. Reston, VA: USGS.
[49] USGS (U.S. Geological Survey). 2008f. *USGS Minerals Yearbook Volume I: Cobalt chapter*. Reston, VA: USGS.
[50] USGS (U.S. Geological Survey). (2008g). *USGS Minerals Yearbook Volume I: Gallium chapter*. Reston, VA: USGS.
[51] USGS (U.S. Geological Survey). (2010f). *USGS Minerals Yearbook Volume I: Cobalt chapter*. Reston, VA: USGS.
[52] USGS (U.S. Geological Survey). (2010g). Historical statistics for mineral and material commodities in the United States. Reston, VA: USGS.
[53] USGS (U.S. Geological Survey). (2010h). *2010 Mineral Commodity Summary: Bauxite and Alumina*. Reston, VA: USGS.
[54] USGS (U.S. Geological Survey). (2010i). "Cobalt Slides," presented to the U.S. Department of Energy, May, 2010.
[55] USGS (U.S. Geological Survey). (2010j). Table of Rare Earth Elemental Breakdown

by Select Sites, information provided by USGS to the Department of Energy, September, 2010.

[56] W.R. Grace & Co. (2010). "Grace Announces Rare Earth Surcharge for FCC Catalysts and Additives." August 25. http://www.grace.com/Media/NewsItem.aspx?id=1463179.

[57] China Daily. (2009). "China Mulls Plans to Curb Rare Earth Smuggling." September 14. http://www.chinadaily.com.cn/bizchina/2009-09/14/content_8690293.htm.

[58] OSD (Office of the Secretary of Defense). (2009). *Reconfiguration of the National Defense Stockpile Report to Congress*. Washington, DC: U.S. Department of Defense

[59] AP (Associated Press). (2010). "Nations Wary of Dependence on China's Rare Earths." *Associated Press*, October 4.

[60] Bae, J. (2010). "*Strategies and Perspectives for Securing Rare Metals in Korea*." Discussion white paper.

[61] Behre Dolbear. (2010). *Ranking Of Countries for Mining Investment*. s.l.: Behre Dolbear.

[62] "China May Launch Rare Earth Strategic Reserve Mechanism Next Year – Source." (2010). *Business China*, October 27. Accessed November 3. http://en.21cbh.com/HTML/2010-10- 27/xN MDAwMDIwMzIxNA.html

[63] Commission of the European Communities. (2008). *The Raw Materials Initiative—Meeting our Critical Needs for Growth and Jobs in Europe*. Brussels: Commission of the European Communities.

[64] European Commission, Enterprise and Industry. (2010). *Critical raw materials for the EU*. s.l.: Fraunhofer ISI.

[65] Han, J. (2010). "Rare Metal Research Center Opens." *Korea Times*, January 28.

[66] Ha-won, J. (2010). "Kores Sees Golden Opportunity to Increase Mine Holdings." *Seoul Joongang Daily,* June 4.

[67] "Japan Looks Past China for Metals." (2010). *Asahi Shimbun*, October 1.

[68] JOGMEC (Japan Oil, Gas and Metals National Corporation). (2007). "JOGMEC's Activities." Accessed June 21, 2010. http://www.jogmec.go.jp/english/activities/index.html.

[69] Kawamoto, H. (2008). "Japan's Policies to be adopted on Rare Metal Resources." *Science & Technology Trends*, April 2008.

[70] METI (Ministry of Economy, Trade, and Industry). (2008). "Guidelines for Securing Natural Resources." Accessed July 10, 2010. http://www.meti.go.jp/english/press/data/nBackIssue20080328_05.html.

[71] METI (Ministry of Economy, Trade, and and Industry). (2009). Announcement of 'Strategy for Ensuring Stable Supplies of Rare Metals'." Accessed July 10, 2010. http://www.meti.go.jp/english/press/data/20090728_01.html.

[72] Muyuan, C. (2010). "Applications for Rare Earth Mining Suspended." *China Daily*, March 16.
M2i (Materials Innovation Institute). 2009. *Material Scarcity*. Delft, The Netherlands: M2i.

[73] Natural Resources Canada. (1996). The Minerals and Metals Policy of the Government of Canada: Partnerships for Sustainable Development. Ottawa, Canada: Government of Canada.

[74] OSD (United States Office of the Secretary of Defense). (2009). *Reconfiguration of the National Defense Stockpile Report to Congress*. Washington, DC: U.S. Department of

Defense.
[75] Smith, P. (2010). "Australia plans 40% tax on mining sector." *Financial Times,* May 2.
[76] Tabuchi, H. (2010). "Japan Recycles Minerals from Used Electronics," *New York Times*, October 4, 2010.
[77] USGS (United States Geological Survey). (2008). *2008 Minerals Yearbook: Australia.* Reston, VA: U.S. Department of the Interior.
[78] Yan, Z. & Yijun, Y. (2010). "Rare Earth Reserve Approved." *China Daily*, February 11.
[79] Yang, H. 2010. "Posco Acquires Chinese Rare Earth Firm." *China Daily*, June 11.
[80] IEA (International Energy Agency). (2009). *World Energy Outlook 2009.*
[81] IEA (International Energy Agency). (2010a). *Energy Technology Perspectives 2010.*
[82] IEA (International Energy Agency). (2010b). *Phasing Out Incandescent Lamps.*
[83] Indium Corp. (2010). "*Indium, Gallium & Germanium Supply & Price Outlook.*" Presented by Gregory Evans to the TREM 2010 Conference, Washington, DC, April.
[84] Roskill. (2010). "*Rare Earths: An Evaluation of Current and Future Supply.*" presentation by Judith Chegwidden to the 2010 Rare Earth Summit, Beijing, August 4–5.
[85] USGS (U.S. Geological Survey). (2008a). *USGS Minerals Yearbook 2008*: *Indium.* Reston, VA: USGS.
[86] USGS (U.S. Geological Survey). (2008b). *USGS Minerals Yearbook 2008*: *Bauxite and Alumina.* Reston, VA: USGS.
[87] USGS (U.S. Geological Survey). (2008c). *USGS Minerals Yearbook 2008*: Tellurium. Reston, VA: USGS.
[88] USGS (U.S. Geological Survey). 2008d. *USGS Minerals Yearbook 2008*: Cobalt. Reston, VA: USGS.
[89] USGS (U.S. Geological Survey). 2008e. *USGS Minerals Yearbook 2008*: Lithium. Reston, VA: USGS.
[90] USGS (U.S. Geological Survey). (2010). Table of Rare Earth Elemental Breakdown by Select Sites, information provided by USGS to the Department of Energy, September, 2010.
[91] Watts. (2010). "Shock & Ore: The Scramble for Rare Earth Self-sufficiency," *Industrial Minerals.* June. http://www.tasmanmetals.com/i/pdf/IndustrialMinerals Article-June2010.pdf.
[92] NAS (National Academy of Sciences). (2008). *Minerals, Critical Minerals and the U.S. Economy.*
[93] Ames Laboratory. (2010). "*Responses to DOE Request for Information.*" June 7.
[94] Behre Dolbea. (2010). Ranking of Countries for Mining Investment.
[95] Hamilton, T. "Lithium Battery Recycling Gets a Boost." *Technology Review.* Accessed August 12. http://www.technologyreview.com/energy
[96] LaMonica, M. "E-Waste Showdown Unearths Deeper Questions." *Green Tech*, Jan. 16. http://news.cnet.com/8301-11128_3-10435193-54.html.
[97] Matthews, R. G. "Permits Drag on U.S. Mining Projects." *Wall Street Journal,* February 8, 2010.
[98] NRC. 2008. "*Minerals, Critical Minerals the U.S. Economy.*" The National Academies Press.
[99] Umicore. 2010. "*Responses to DOE Request for Information.*" June 7.

[100] Aston, A. (2010). "China's Rare-Earth Monopoly: The rest of the world is trying to find alternatives to these crucial materials." *Technology Review*, October 15. http://www.technologyreview.com/energy

[101] DOE (U.S. Department of Energy) EERE (Energy Efficiency and Renewable Energy) (2009). "2010 Toyota Prius-6063 Hybrid BOT Battery Test Results." U.S. Department of Energy. July 15. http://avt.inel.gov/pdf/hev/batterygenIIIprius6063.pdf.

[102] DOE (U.S. Department of Energy) EERE (Energy Efficiency and Renewable Energy). (2010). "*Expert judgment by vehicle battery specialist.*"

[103] Electron Energy Corporation. (2010). "*Response to U.S. Department of Energy Request for Information.*" June 5.

[104] Gains, L. & Nelson, P. (2009). "*Lithium-Ion Batteries: Possible Material Demand Issues.*" Argonne National Laboratory. http://www.transportation

[105] General Electric Corporation. (2010). "*Response to U.S. Department of Energy Request for Information.*" June 7.

[106] Grana, P. (2010). "*The Efficiency Race: How do CdTe and CIGS Stack Up?*" Renewableenergyworld.com. http://www.renewableenergyworld.com/rea/blog/post/2010/11/theefficiency-race

[107] International Energy Agency (IEA). (2010). *Information Paper: Phase Out of Incandescent Lamps*. http://www.iea.org/papers/2010/phase_out.pdf.

[108] Lifton, J. (2009). "The Rare Earth Crisis of 2009." *The Jack Lifton Report*. October 15. In Oakdene Hollins 2010.

[109] Lynas Corporation. (2010). "Response to U.S. Department of Energy Request for Information." June 7.

[110] National Research Council. (2010). "Transitions to Alternative Transportation Technologies – Plug-in Hybrid Electric Vehicles." http://www.nap.edu/catalog.php?record_id=12826.

[111] NREL (National Renewable Energy Laboratory). (2010). Email communication, September 17, 2010.

[112] Oakdene Hollins. 2010. "*Response to U.S. Department of Energy Request for Information.*" May 26.

[113] Reddy, Thomas R. ed. 2011. *Linden's Handbook of Batteries*. New York: McGraw Hill.

[114] Toyota. (2009). "*Toyota Prius Emergency Response Manual, Hybrid 2010 Model 3rd Generation.*" https://techinfo.toyota.com/techInfoPortal/staticcontent/en/techinfo/html/prelogin/docs/3rdprius. pdf

End Notes

[1] Samarium cobalt (SmCo) rare earth PMs are used for certain niche applications, particularly in the defense sector. They are slightly less powerful by size and weight than NdFeB magnets, although they continue to operate effectively in higher temperatures (Electron Energy Corporation 2010). This high temperature operating capability makes SmCo magnets the appropriate choice for some applications.

[2] Magnequench merged with Canadian based AMR Technologies in 2005 to form Neo Materials Technologies. It now operates as a Canadian-based company with Chinese operations.

[3] Lithium titanate batteries use a lithium titanium oxide anode and have been mentioned as a potential candidate for automotive use (Gains 2010), despite being limited by a low cell voltage compared to other Li- ion battery chemistries. Lithium titanate is used later in the strategy as the lithium-ion battery chemistry with the highest lithium content in creating material demand projections.

[4] Concentrating PV and nanotechnology-based films remain a negligible share of the market.

[5] In this Strategy, we use tonnes rather than metric tons with which the reader might be more familiar. 1 tonne = 1 metric ton (Mt).

[6] NREL, Email communication, September 17, 2010.

[7] This discussion of lighting technologies does not consider niche lighting technologies such as high-intensity discharge (HID) lamps, which are very efficient but are used almost exclusively outdoors or in very large indoor areas.

[8] Much higher purity than for the other REE applications described earlier.

[9] Calculation assuming feed rates of 0.21 to 0.25 pounds of catalysts per barrel FCC feed and 2% REO content.

[10] Data in this table are from the most recent data available from USGS.

[11] Approximately 20,000 additional REO from "unofficial" sources (Kingsnorth and Chegwidden 2010).

[12] This 2009 production figure is from Molycorp (2010).

[13] Cobalt Development Institute (2009).

[14] This set of data on indium based on indium content is from 2008 (USGS 2008d); the breakdown of "Others" is not available.

[15] Others include Australia, Belgium, China, France, Germany, Kazakhstan, the Philippines, Russia and the UK (USGS 2010f).

[16] Mine production information for Ga is production capacity in 2008 rather than production in 2009 (USGS 2008g).

[17] USGS, telephone communication, December 7, 2010.

[18] Only part of the gallium present in bauxite and zinc ores is recoverable with existing technology, and the factors controlling the recovery are proprietary. An estimate of current reserves of gallium comparable to the definition of reserves of other minerals thus cannot be made (USGS 2010c).

[19] Trace amounts (<1%) of heavy rare earths are also found in monazite mineral, except for yttrium, whose abundance in monazite is higher (up to 2.5% in currently known projects) (USGS 2010, Roskill 2010, IMCOA 2010). A slightly larger concentration of heavy rare earths is also found in the fergusonite deposit at one of the mines (Nechalaco) likely to come online in the next five years (Roskill 2010).

[20] USGS, in-person meeting and multiple telephone communication, July–September, 2010.

[21] Sum of concentrations may not total 100% due to matrix effect when analyzing various natural materials that may differ in composition from the control standards used in calibration. Chart modified from *USGS Minerals Yearbook 2007 Volume I:* Rare Earths chapter, Table 2, p. 60.11.

[22] Currently the Mountain Pass site is limited to the reprocessing of rare earth ores from previously mined stocks.

[23] Monazite production at this Brazil site is unknown (General Electric 2010).

[24] Ore, rather than the normally quoted mineral (Roskill 2010).

[25] The paucity of data is also due to a lack of mineralogical studies or mine plan development for some of the rare earth deposits.

[26] Other prominent sources of rare earth data include industry consultants (e.g., Roskill Information Services Ltd or Roskill, Industrial Minerals Company of Australia Pty Ltd or IMCOA, the Anchor House), major mining firms, and General Electric.

[27] India has reported an almost unchanging production level of REEs of 2,500-2,700 tonnes since 1994, and is currently second to China globally in rare earth production. Brazil, which saw a rise in production in the late 1990s, has been producing in the 650 tonnes range. Malaysia has been producing around 380 tonnes per year (USGS 1994-2010).

[28] The production capacity outside of China for 2010 is 10,000–12,000 tonnes at best, indicating a shortfall in 2010 of at least 10,000–15,000 tonnes (Hatch 2010).

[29] In this Strategy, we use tonnes rather than metric tons with which the reader might be more familiar. 1 tonne = 1 metric ton (Mt).

[30] China issues export licenses for rare earths twice a year.

[31] The production from the Commonwealth of Independent States is not available between 2005 and 2009, and that from other countries not available between 2006 and 2009 according USGS data. The 2010 production capacity outside China is estimated based on the 10,000–12,000 tonnes shortfall predicted by Hatch (2010).

[32] Chile's estimated reserve volume is at 7.5 million tonnes (USGS 2010a).

[33] USGS 1994-2010a.

[34] USGS, external review of earlier draft, November 17, 2010.

[35] The global production does not include U.S. production data, information withheld by the USGS to avoid disclosure of proprietary information (USGS 2010a).

[36] These shares are based on global production not including U.S. production (see the previous footnote).

[37] China is the only country in the world still converting lithium minerals into compounds from spodumene or hard rocks, including imported lithium from spodumene in Australia (USGS, correspondence, August 24 2010). In fact, its annual domestic production of lithium minerals from hard rock ores has been rising by approximately 15% per year since 2000 (Roskill 2009).

[38] USGS, 1994-2010a.

[39] Australia holds the next largest reserves or approximately 23% of the worldwide total. The United States possesses an estimated 33,000 tons of cobalt reserves or around 1% global reserves (USGS 2010b).

[40] Cobalt production from the DRC occurs in the Copperbelt of Katanga Province, not in the conflict areas of North and South Kivu provinces (USGS correspondence 10/25).

[41] The Policy Potential Index (PPI), generated annually by the Canadian Frasier Institute, is based on a survey of investors of mining ventures. It gauges the extent to which countries are putting up social and political barriers to entry. The index takes into account public policy factors such as taxation and regulation affecting exploration investment (McMahon and Cervantes 2010).

[42] USGS, telephone communication, October 25 2010.

[43] USGS, telephone communication, October 25, 2010.

[44] USGS 1994-2010b.

[45] Also based on multiple email exchanges and phone communication with USGS, October 4-7, 2010.

[46] USGS 1994-2010d.

[47] Generally and with respect to the key materials, demand for end-use items for building use (e.g., phosphors for lighting) or construction tend to be more cyclical, whereas those that enter big-ticket consumer items such as cars tend to be more volatile and sensitive to short-term economic movements. Uses that enter portable devices and personal consumer goods (e.g., batteries for portable electronics) tend to experience more stable demand (Humphreys, forthcoming). Regional factors are important also: China's and India's rapid economic growth have had and will continue to have a huge impact on global demand for mineral commodities (Eggert 2010).

[48] Import dependence by itself need not be considered a risk factor. Rather, the possibility of supply disruptions is due to a combination of heavy import reliance and concentration of supply in a few companies or countries that may be unreliable suppliers.

[49] End use shares are from the most recent USGS data available; the shares are global for Li and Co; and domestic for REEs, In, Te, and Ga. The data source for Te is *Umicore* (2010).

[50] Based on information provided by USGS on September 14, 2010.

[51] Canada, UK, US and Taiwan are involved in gallium recycling (USGS, correspondence, December 7, 2010).

[52] Estimated share of U.S. gallium consumption out of global total is based on the 2008 global production number instead as global demand data for gallium is harder to determine and it is reasonable to assume that global Ga supply is at a similar level as global Ga demand; data source is USGS (correspondence 10/19).

[53] Demand here refers to demand by manufacturers of materials for production rather than demand by households for final products containing the material.

[54] China's total annual rare earth consumption in these sectors has grown from a mere 1% in 1987 to 60% in 2008 (Chen 2010).

[55] USGS 1994-2010.

[56] Global end-use markets are estimated as follows: ceramics and glass, 31%; batteries, 23%; lubricating greases, 10%; air treatment, 5%; continuous casting, 4%; primary aluminum production, 3%; and other uses, 24% (USGS 2010a).

[57] The United States has been re-exporting around half of its lithium imports since 2003 (USGS1994–2010c).

[58] USGS 1994-2010a.

[59] Optoelectronic devices were used in aerospace, consumer goods, industrial equipment, medical equipment and telecommunications. ICs were used in defense applications, high-performance computers and telecommunications (USGS 2010c).

[60] Gallium stocks have met an average of 8% of annual demand in the United States since 1982 (USGS 1994–2010c).

[61] The United States stopped exporting gallium in 1984 (USGS 1994-2010c).

[62] The United States had minimal stocks of indium from 1993 to 1998 and exported an average of 19% of its annual imports between 1999 and 2002 (USGS 1994-2010d).

[63] For example, in August 2010 W.R. Grace established a rare earth surcharge that increased the price of its fluid cracking catalysts and additives due to rapidly increasing rare earth prices (W.R. Grace 2010).

[64] Examples cited in Humphreys (forthcoming 2011) include the power shortage concerns in Southern Africa and Chile where mining is important; water, which is required in large quantities by mining and is becoming scarcer and more expensive in some parts of the world; a move toward smaller and higher cost deposits and resources; and increasing political barriers to entry.

[65] In China, gallium is also traded through informal metals exchanges where transparent pricing and a spot market are present.

[66] USGS, external review of earlier draft, November 17, 2010.

[67] In 2004, world demand for cobalt reportedly increased as a result of an increase in demand from the aerospace and land-based gas turbine industries and growth of cobalt use in rechargeable batteries and catalysts (USGS 2004a).

[68] This period led up to the European Union's (EU) enactment of a new chemical policy known as "REACH" which affected all suppliers of cobalt as well as other materials to the European market by requiring them to collect and submit risk assessment data on each material produced in or imported to the EU. The goals of REACH included "Improved protection of human health and the environment." http://guidance

[69] USGS 1994-2010b.

[70] Total global mine production of lithium materials was 6,100 tonnes in 1994 (USGS 1996a); U.S. production info which was withheld to protect proprietary information is not reflected in the global number.

[71] USGS 1994-2010a

[72] USGS 1994-2010c.

[73] Some historical data and basic growth projections exist for non-clean energy applications; however, future demand will be governed by complex market dynamics, including the extent to which increased clean energy demand prompts innovations that reduce material requirements for all applications.

[74] Vehicles with electric drive-trains include hybrid gasoline-electric and diesel-electric vehicles, plug-in hybrid electric vehicles and all-electric vehicles.

[75] Data sources: estimated 2010 production based on personal communication with Kingsnorth, Roskill (2010) for assumed additional production by 2015 with downward adjustments to Mt. Weld and Nolans Bore to reflect expected supply by 2015, elemental breakdowns based on USGS (2010).

[76] The mines and assumed respective supplies are (i) North Queensland Metals Baal Gamm mine in Australia (15 tonnes/yr), (ii) South American Silver's Malku Khota mine in Bolivia (15–20 tonnes/yr), (iii) Votorantim Metais' Juiz de For a mine in Brazil (15 tonnes/yr) and (iv) UMMC's Electrozinc facility in Russia (2 tonnes/yr).

[77] For indium, the additional amount is only the difference between the 2010 production and the maximum current production capacity for mining and refining the material. No new capacity is projected by 2015.

[78] Also based on multiple correspondences with USGS, October 4-7, 2010.

[79] USGS, external review of earlier draft, November 17, 2010.

[80] Currently and for the foreseeable future, Bolivia's lithium is only an uneconomic resource. It is unknown if Boliva will ever be able to turn its lithium resource into an economic reserve (USGS, pers. comm.).

[81] The substitution of praseodymium for neodymium is not considered in the demand trajectories due to a lack of available data on the extent to which it occurs.

[82] Because REEs used in lighting need to be very pure (99.999%), the rare earth oxides (REOs) sold to phosphor manufacturers are much more expensive than those used by manufacturers of other REE applications. In the event of a material shortage, producers of REOs would likely divert the available supply of a given element into phosphors rather than the other applications of this element due to the greater profit margins. Therefore, the impact of shortages in overall REE supplies may have a limited effect on the availability of lighting phosphors (Gschneidner, pers. comm.).

[83] Estimated clean energy demand in 2010 slightly exceeds data for total global demand. This is likely explained by data uncertainty about total global demand and the dependency on a single set of phosphor content assumptions. It is clear that non-clean energy demand is greater than zero currently and in the future.

[84] The additional supply may come from existing stockpiles or may be due to data uncertainty.

[85] EPA, for example, has delegated federal regulatory authorities to states under the Clean Air Act, Clean Water Act and Resource Conservation and Recovery Act. A mine or processing plant may be required to obtain a series of permits under these acts before such a facility can begin operations. The securing of such permits requires the submittal of complex environmental modeling results which must show that a facility can meet its regulatory standards during operations. Further, if a mine or processing plant triggers a "major federal action" as defined under the National Environmental Protection Act, the facility may also be required to prepare an Environmental Impact Statement (EIS). Such an EIS can take up to two to five years to complete.

[86] The Council for Environmental Quality is leading an interagency group for providing rapid response capability to fix coordination problems between federal agencies on renewable and transmission permit applications. A similar approach could apply to the mining permit process as well.

[87] DOE also has limited authorities under the Defense Production Act (DPA) that could affect the production of critical materials. Under Title I of the DPA, DOE can exercise priority rights over private parties in procurement for projects that maximize domestic energy supplies, including those that maintain or further domestic energy exploration, production and refining. Title III of the DPA authorizes the President to provide incentives including loan guarantees for the development, modernization or expansion of domestic defensive productive capacity and supply. Recent Title III projects have included the development of a U.S.-owned domestic source for prismatic lithium-ion cells and batteries for spacecraft use.

[88] The DoD defines a "strategic material" as one 1) which is essential for important defense systems, 2) which is unique in the function it performs and 3) for which there are no viable alternatives. A "critical material" is a strategic material for which 1) the Department of Defense dominates the market for the material, 2) the

Department's full and active involvement and support are necessary to sustain and shape the strategic direction of the market and 3) there is significant and unacceptable risk of supply disruption due to vulnerable U.S. or qualified non-U.S. suppliers.

In: Critical Materials Strategy for Clean Energy Technologies ISBN: 978-1-61209-829-6
Editor: Nicholas J. Bartlett

Chapter 2

RARE EARTH ELEMENTS: THE GLOBAL SUPPLY CHAIN

Marc Humphries

SUMMARY

The concentration of production of rare earth elements (REEs) outside the United States raises the important issue of supply vulnerability. REEs are used for new energy technologies and national security applications. Is the United States vulnerable to supply disruptions of REEs? Are these elements essential to U.S. national security and economic well-being?

There are 17 rare earth elements (REEs), 15 within the chemical group called lanthanides, plus yttrium and scandium. The lanthanides consist of the following: lanthanum, cerium, praseodymium, neodymium, promethium, samarium, europium, gadolinium, terbium, dysprosium, holmium, erbium, thulium, ytterbium, and lutetium. Rare earths are moderately abundant in the earth's crust, some even more abundant than copper, lead, gold, and platinum. While more abundant than many other minerals, REE are not concentrated enough to make them easily exploitable economically. The United States was once self-reliant in domestically produced REEs, but over the past 15 years has become 100% reliant on imports, primarily from China, because of lower-cost operations.

There is no rare earth mine production in the United States. U.S.-based Molycorp operates a separation plant at Mountain Pass, CA, and sells the rare earth concentrates and refined products from previously mined above-ground stocks. Neodymium, praseodymium, and lanthanum oxides are produced for further processing but these materials are not turned into rare earth metal in the United States.

Some of the major end uses for rare earth elements include use in automotive catalytic converters, fluid cracking catalysts in petroleum refining, phosphors in color television and flat panel displays (cell phones, portable DVDs, and laptops), permanent magnets and rechargeable batteries for hybrid and electric vehicles, and generators for wind turbines, and numerous medical devices. There are important defense applications, such as jet fighter engines, missile guidance systems, antimissile defense, and space-based satellites and communication systems.

World demand for rare earth elements is estimated at 134,000 tons per year, with global production around 124,000 tons annually. The difference is covered by previously mined aboveground stocks. World demand is projected to rise to 180,000 tons annually by 2012, while it is unlikely that new mine output will close the gap in the short term. New mining projects could easily take 10 years to reach production. In the long run, however, the USGS expects that global reserves and undiscovered resources are large enough to meet demand.

Legislative proposals **H.R. 6160** (Dahlkemper) **H.R. 4866** (Coffman) and **S. 3521**(Murkowski) have been introduced to support domestic production of REEs, because of congressional concerns over access to rare earth raw materials and downstream products used in many national security applications and clean energy technologies. The House approved H.R. 6160 on September 29, 2010, by a vote of 325-98.

INTRODUCTION

The concentration of production of rare earth elements (REEs) raises the important issue of supply vulnerability. REEs are used for new energy technologies and national security applications. Is the U.S. vulnerable to supply disruptions? Are these elements essential to U.S. national security and economic well-being?

The examination of REEs for new energy technologies reveals a concentrated and complex global supply chain and numerous end-use applications. Placing the REE supply chain in the global context is unavoidable. U.S. mineral policy emphasizes developing domestic supplies of critical materials and the domestic private sector to produce and process those materials. [1] But some raw materials do not exist in economic quantities in the United States, while processing, manufacturing, and other downstream ventures in the United States may not be competitive with facilities in other regions of the world. However, there may be public policies enacted or executive branch measures taken to offset the U.S. disadvantage of its potentially higher cost operations. The current goal of U.S. mineral policy is to promote an adequate, stable, and reliable supply of materials for U.S. national security, economic well-being, and industrial production.

Aside from a small amount of recycling, the United States is 100% reliant on imports of REEs and highly dependent on many other minerals that support its economy. For example, the United States is more than 90% import-reliant for manganese (100%), bauxite (100%), platinum (94%), and uranium (90%). While import reliance may be a cause for concern, high import reliance is not necessarily the best measure, or even a good measure, of supply risk. The supply risk for bauxite, for example, may not be the same as that for REEs. However, in the case of REEs, the dominance of China as a single or dominant supplier of the raw material, downstream oxides, associated metals and alloys, is a cause for concern because of China's growing internal demand for its REEs.

This report provides a discussion on the major issues and concerns of the global supply chain for rare earth elements, their major end uses, and legislative and other policy proposals that Congress may consider to improve the U.S. rare earth position.

WHAT ARE RARE EARTH ELEMENTS?

There are 17 rare earth elements (REEs), 15 within the chemical group called lanthanides, plus yttrium and scandium. The lanthanides consist of the following: lanthanum, cerium, praseodymium, neodymium, promethium, samarium, europium, gadolinium, terbium, dysprosium, holmium, erbium, thulium, ytterbium, and lutetium. Rare earths are moderately abundant in the earth's crust, some even more abundant than copper, lead, gold, and platinum. While some are more abundant than many other minerals, most REEs are not concentrated enough to make them easily exploitable economically.[2] The United States was once self-reliant in domestically produced REEs, but over the past 15 years has become 100% reliant on imports, primarily from China, because of lower-cost operations.[3]

MAJOR END USES AND APPLICATIONS

Currently, the dominant end use for rare earth elements in the U.S. are for auto catalysts and petroleum refining catalysts. Some other major end uses for rare earth elements include use in phosphors in color television and flat panel displays (cell phones, portable DVDs, and laptops), permanent magnets and rechargeable batteries for hybrid and electric vehicles, and numerous medical devices. There are important defense applications such as jet fighter engines, missile guidance systems, antimissile defense, and space-based satellites and communication systems. Permanent magnets containing neodymium, gadolinium, dysprosium, and terbium are used in numerous electrical and electronic components and generators for wind turbines. See **Table 1** below for selected end uses of rare earth elements.

Table 1. Rare Earth Elements (Lanthanides): Selected End Uses

Light Rare Earths (more abundant)	Major End Use	Heavy Rare Earth (less abundant)	Major End Use
Lanthanum	hybrid engines, metal alloys	Terbium	phosphors, permanent magnets
Cerium	auto catalyst, petroleum refining, metal alloys	Dysprosium	permanent magnets, hybrid engines
Praseodymium	magnets	Erbium	phosphors
Neodymium	auto catalyst, petroleum refining, hard drives in laptops, headphones, hybrid engines	Yttrium	red color, fluorescent lamps, ceramics, metal alloy agent
Samarium	magnets	Holmium	glass coloring, lasers
Europium	red color for television and computer screens	Thulium	medical x-ray units
Gadolinium	magnets	Lutetium	catalysts in petroleum refining
		Ytterbium	lasers, steel alloys

Source: DOI, U.S. Geological Survey, Circular 930-N.

Demand for Rare Earth Elements

World demand for rare earth elements is estimated at 134,000 tons per year, with global production around 124,000 tons annually. The difference is covered by above-ground stocks or inventories. World demand is projected to rise to 180,000 tons annually by 2012, while it is unlikely that new mine output will close the gap in the short term.[4] By 2014, global demand for rare earth elements may exceed 200,000 tons per year. China's output may reach 160,000 tons per year (up from 130,000 tons in 2008) in 2014. An additional capacity shortfall of 40,000 tons per year may occur. This potential shortfall has raised concerns in the U.S. Congress. New mining projects could easily take 10 years for development. In the long run, however, the USGS expects that reserves and undiscovered resources are large enough to meet demand.

While world demand continues to climb, U.S. demand for rare earths is also projected to rise, according to the USGS Commodity Specialist Jim Hedrick.[5] For example, permanent magnet demand is expected to grow by 10%-16% per year through 2012. Demand for rare earths in auto catalysts and petroleum cracking catalysts is expected to increase between 6% and 8% each year over the same period. Demand increases are also expected for rare earths in flat panel displays, hybrid vehicle engines, and defense and medical applications.

The Application of Rare Earth Metals in National Defense[6]

Current government policies pertaining to the acquisition of certain minerals for defense purposes are addressed, in part, in several different legislative initiatives, including the Defense Production Act (P.L. 8 1-774), National Defense Stockpile [Title 50 United States Code (U.S.C.) 98-h-2(a)],[7] Buy American Act (41 U.S.C. 10-10d), Berry Amendment (10 U.S.C. 2533a), and the Specialty Metal provision (10 U.S.C. 2533b). However, these policies do not present a unified opinion on whether every mineral is considered "critical," "strategic," or necessary for national security purposes, and there is a certain lack of cohesion to the application of these policies. As an example, rare earth elements (and rare earth metals) fall outside of the scope of the Berry Amendment and the Specialty Metal provision.[8]

The primary defense application of rare earth materials is their use in four types of permanent magnet materials commercially available: Alnico, Ferrites, Samarium Cobalt, and Neodymium Iron Boron. With the exception of Neodymium Iron Boron, all of the materials are domestically produced. The United States has no production capabilities for Neodymium Iron Boron. Neo magnets, the product derived from Neodymium Iron Boron, and Samarium Cobalt, are considered important to many defense products. They are considered one of the world's strongest permanent magnets and an essential element to many military weapons systems, as described in the following examples.

- Jet fighter engines and other aircraft components, including samarium-cobalt magnets used in generators that produce electricity for aircraft electrical systems;
- Missile guidance systems, including precision guidance munitions, lasers, and smart bombs;[9]
- Electronic countermeasures systems;

- Underwater mine detection systems;
- Antimissile defense systems;
- Range finders, including lasers; and
- Satellite power and communication systems, including traveling wave tubes (TWT) rare earth speakers, defense system control panels, radar systems, electronic counter measures, and optical equipment.[10]

Many scientific organizations have concluded that certain rare earth metals are critical to U.S. national security and becoming increasingly more important in defense applications.[11] Some industry analysts are concerned with an increasing dependence on foreign sources for rare earth metals; a dwindling source of domestic supply for certain rare earth metals; and the emergence of a manufacturing supply chain that has largely migrated outside of the United States. In July 2010, the China Ministry of Commerce announced that China would cut its export quota for rare earth minerals by 72%, raising concerns because of estimates that China controls approximately 97% of the global production of rare earth minerals.[12] It is also estimated that by 2012 China's domestic consumption will outpace China's domestic production of rare earth minerals.

Some experts are concerned that DOD is not doing enough to mitigate the possible risk posed by a scarcity of domestic suppliers. As an example, the United States Magnet Materials Association (USMMA), a coalition of companies representing aerospace, medical, and electronic materials, has recently expanded its focus to include rare earth metals and the rare earth magnet supply chain. In February 2010, USMMA unveiled a six-point plan to address what they describe as the "impending rare earth crisis" which they assert poses a significant threat to the economy and national security of the United States.[13] However, it appears that DOD's position assumes that there are a sufficient number of supplier countries worldwide to mitigate the potential for shortages.

Rare Earth Resources and Production Potential

Rare earth elements often occur with other elements, such as copper, gold, uranium, phosphates, and iron, and have often been produced as a byproduct. The lighter elements such as lathananum, cerium, praseodymium, and neodymium are more abundant and concentrated and usually make up about 80%-99% of a total deposit. The heavier elements—gadolinium through lutetium and yttrium—are scarcer but very "desirable," according to USGS commodity analysts.[14]

Most rare earth elements throughout the world are located in deposits of the minerals bastnaesite[15] and monazite.[16] Bastnaesite deposits in the United States and China account for the largest concentrations of REEs, while monazite deposits in Australia, South Africa, China, Brazil, Malaysia, and India account for the second largest concentrations of REEs. Bastnaesite occurs as a primary mineral, while monazite is found in primary deposits of other ores and typically recovered as a byproduct. Over 90% of the world's economically recoverable rare earth elements are found in primary mineral deposits (i.e., in bastnaesite ores).[17]

Concerns over radioactive hazards associated with monazites (because it contains thorium) has nearly eliminated it as a REE source in the United States. Bastnaesite, a low-thorium mineral (dominated by lanthanum, cerium, and neodymium) is shipped from stocks in Mountain Pass, CA. The more desirable heavy rare earth elements account for only 0.4% of the total stock. Monazites have been produced as a minor byproduct of uranium and niobium processing. Rare earth element reserves and resources are found in Colorado, Idaho, Montana, Missouri, Utah, and Wyoming. Heavy rare earth elements (HREEs) dominate in the Quebec-Labrador (Strange Lake) and Northwest Territories (Thor Lake) areas of Canada. There are high-grade deposits in Banyan Obo, Inner Mongolia, China (where much of the world's REE production is taking place) and lower-grade deposits in South China provinces providing a major source of the heavy rare earth elements.[18] Areas considered to be attractive for REE development include Strange Lake and Thor Lake in Canada; Karonga, Burundi; and Wigu Hill in Southern Tanzania.

Careful consideration should be given to the feasibility of mining and processing of REEs as a byproduct of phosphorus deposits and from titanium and niobium mines in Brazil.[19] Canadian, Chinese, and U.S. firms have recently assessed various REE deposits associated with development of primary minerals such as gold, iron ore, and mineral sand projects in the United States. **Table 2** below illustrates China's near-monopoly position in world rare earth production. However, REE reserves and the reserve base are more dispersed throughout the world. China holds 36% of the world's reserves (36 million metric tons out of 99 million metric tons) and the United States holds about 13%. South Africa and Canada (included in the "Other" category) have significant REE potential, according to the USGS. REE reserves are also found in Australia, Brazil, India, Russia, South Africa, Malaysia, and Malawi.

Table 2. Rare Earth Elements: World Production and Reserves—2009

Country	Mine Production (metric tons)	% of total	Reserves (million metric tons)	% of total	Reserve Basea (million metric tons)	% of total
United States	none		13.0	13	14.0	9.3
China	120,000	97	36.0	36	89.0	59.3
Russia(and other former Soviet Union countries)			19.0	19	21.0	14
Australia			5.4	5	5.8	3.9
India	2,700	2	3.1	3	1.3	1
Brazil	650		small			
Malaysia	380		small			
Other	270		22.0	22	23	12.5
Total	124,000		99.0		154	

Source: U.S. Department of the Interior, Mineral Commodity Summaries, USGS, 2010.

a. Reserve Base is defined by the USGS to include reserves (both economic and marginally economic) plus some subeconomic resources (i.e., those that may have potential for becoming economic reserves).

There is no rare earth mine production in the United States. U.S.-based Molycorp operates a separation plant at Mountain Pass, CA, and sells the rare earth concentrates and refined products from previously mined above-ground stocks. Neodymium, praseodymium, and lanthanum oxides are produced for further processing, but these materials are not turned into rare earth metal in the United States. While the United States exports much of its REE stocks to Japan, that material is not counted in the trade equation for import reliance because the material is not produced from a primary source.

Molycorp, which has an exploration program underway to further delineate its rare earth mineral deposits, has plans for full mine production in the second half of 2012 and has plans to modernize its refinery facilities. Molycorp's Mountain Pass deposit contained an estimated 30 million tons of REE reserves and once produced as much as 20,000 tons per day.[20] Mountain Pass cut-off grade (below which the deposit may be uneconomic) is, in some parts, 7.6%, while the average grade is 9.6%. U.S. Rare Earth (another U.S. based company), in the pre-feasibility stage of mine development, has long-term potential because of its large deposits in Idaho, Colorado, and Montana.[21]

Canadian deposits contain the heavy rare earth elements dysprosium, terbium, and europium, which are needed for magnets to operate at high temperatures. Great Western Minerals Group (GWMG) of Canada and Avalon Rare Metals have deposits with an estimated high content (1%- 2%) of heavy rare earth elements.[22] Avalon is developing a rare earth deposit at Thor Lake in the Northwest Territories of Canada. Drilling commenced in January 2010. Thor Lake is considered by some in the industry to contain one of the largest REE deposits in the world with the potential for production of heavy REEs.[23] GWMG owns a magnet alloy producer in the U.K. When GWMG begins production in Canada and elsewhere, they plan to have a refinery near the mine site allowing greater integration and control over the supply chain. Great Western's biggest advantage could be its potential for a vertically integrated operation.

The Lynas Corp., based in Australia, has immediate potential for light rare earths development, according to investor analyst Jack Lifton. Development of Lynas's Mt. Weld deposit in Australia is underway and there is potential to reopen the rare earth mine Steenkampskraal in South Africa. An agreement between GWMG and Rare Earth Extraction Co. Ltd. of Stellenbosch to develop the mine is in progress. The Japan Oil, Gas, and Metals National Corporation (JOGMEC) signed an agreement with Midland Exploration Inc. for development of the Ytterby project in Quebec, Canada. JOGMEC is under the authority of the Japanese Ministry of Economy, Trade, and Industry with a mandate to invest in projects worldwide to receive access to stable supplies of natural resources for Japan.

Access to a reliable supply to meet current and projected demand is an issue of concern. In 2009, China produced 97% of the world's rare earth elements (measured in rare earth oxide content) and continues to restrict exports of the material through quotas and export tariffs. China has plans to reduce mine output, eliminate illegal operations, and restrict REE exports even further. There are some immediate supply concerns with lower rare earth export quotas in China. China has cut its exports of rare earth elements from about 50,000 metric tons in 2009 to 30,000 metric tons in 2010. According to a Bloomberg news report, a July 2010 announcement by China's Ministry of Commerce would cut exports of REEs by 72%, to about 8,000 metric tons, for the second half of 2010.[24]

While limited production and processing capacity for rare earths currently exists elsewhere in the world, additional capacity is expected to be developed in the United States,

Australia, and Canada within two to five years, according to some experts.[25] Chinese producers are also seeking to expand their production capacity in areas around the world, particularly in Australia. There are only a few exploration companies that develop the resource, and because of long lead times needed from discovery to refined elements, supply constraints are likely in the short term.

Supply Chain Issues

The supply chain for rare earth elements generally consist of mining, separation, refining, alloying, and manufacturing (devices and component parts). A major issue for REE development in the United States is the lack of refining, alloying, and fabricating capacity that could process any future rare earth production. There is one U.S. company, Electron Energy Corporation (EEC) in Landisville, PA, producing samarium-cobalt (Sm-Co) permanent magnets, while there are no U.S. producers of the more desirable neodymium-iron-boron (NdFeB) magnets needed for numerous consumer electronics, energy, and defense applications. EEC, in its production of its Sm-Co permanent magnet, uses small amounts of gadolinium—an REE of which there is no U.S. production. Even the REEs needed for these magnets that operate at the highest temperatures include small amounts of dysprosium and terbium, both available only from China at the moment. EEC imports magnet alloys used for its magnet production from China. Santoku America, Inc., produces at its Tolleson, AZ, facility, both NdFeB and Sm-Co alloys used in the production of permanent magnets. They are the only U.S. producer of the NdFeB alloy.

A Government Accountability Office (GAO) report illustrates the lack of U.S. presence in the REE global supply chain at each of the five stages of mining, separation, refining oxides into metal, fabrication of alloys and the manufacturing of magnets and other components.[26] China produces 97% of the REE raw materials, about 97% of rare earth oxides, and is the only exporter of commercial quantities of rare earth metals (Japan produces some metal for its own use for alloys and magnet production). About 90% of the metal alloys are produced in China (small production in the United States) and China manufactures 75% of the neodymium magnets and 60% of the samarium magnets. A small amount of samarium magnets are produced in the United States. Thus, even if rare earth production ramps up, much of the processing/alloying and metal fabrication would occur in China. According to investor analyst Jack Lifton, the rare earth metals are imported from China, then manufactured into military components in the United States or by an allied country.

Lifton states that many investors believe that for financing purposes, it is not enough to develop REE mining operations alone without building the value-added refining, metal production, and alloying capacity that would be needed to manufacture component parts for end-use products. According to Lifton, vertically integrated companies may be more desirable. It may be the only way to secure investor financing for REE production projects.[27] Joint ventures and consortiums could be formed to support production at various stages of the supply chain at optimal locations around the world. Each investor or producer could have equity and offtake commitments. Where U.S. firms and its allies invest is important in meeting the goal of providing a secure and stable supply of REEs, intermediate products, and component parts needed for the assembly of end-use products.

Role of China

State-run ("State-Key") labs in China have consistently been involved in research and development of REEs for over fifty years. There are two State-Key labs: (1) Rare Earth Materials Chemistry and Applications, which has focused on rare earth separation techniques and is affiliated with Peking University, and (2) Rare Earth Resource Utilization, which is associated with the Changchun Institute of Applied Chemistry. Additional labs concentrating on rare earth elements include the Baotou Research Institute of Rare Earths, the largest rare earth research institution in the world, established in 1963, and the General Research Institute for Nonferrous Metals established in 1952.[28] This long term outlook and investment has yielded significant results for China's rare earth industry.

Major iron deposits in the Bayan Obo in Inner Mongolia contain significant rare earth elements recovered as a byproduct or co-product of iron ore mining. China has pursued policies that would use Bayan Obo as the center of rare earth production and R&D. REEs are produced in the following provinces of China: Baotao (Inner Mongolia) Shangdong, Jiangxi, Guangdong, Hunan, Guangxi, Fujian, and Sichuan. Between 1978 and 1989, China's annual production of rare earth elements increased by 40%. Exports rose in the 1990s, driving down prices. In 2007, China had 130 neo-magnet producers with a total capacity of 80,000 tons. Output grew from 2,600 tons in 1996 to 39,000 tons in 2006.

Spurred by economic growth and increased consumer demand, China is ramping up for increased production of wind turbines, consumer electronics, and other sectors, which would require more of its domestic rare earth elements. Safety and environmental issues may eventually increase the costs of operations in China's rare earth industry as domestic consumption is becoming a priority for China. REE manufacturing is set to power China's surging demand for consumer electronics—cell phones, laptops and green energy technologies. According to the report by Hurst, China is anticipating going from 12 gigawatts (GW) of wind energy in 2009 to 100 GW in 2020. Neodymium magnets are needed for this growth.[29]

China's policy initiatives restrict the exports of rare earth raw materials, especially dysprosium, terbium, thulium, lutetium, yttrium, and the heavy rare earths. The export restrictions would not likely affect the downstream metal or magnets. According to Hurst, China wants an expanded and fully integrated REE industry where exports of value-added materials are preferred. It is common for a country to want to develop more value-added production and exports if it is possible. Hurst also suggests China wants to build strategic stockpiles of raw materials as South Korea and Japan have done, and thus have better control over global supply and prices.

The recent (September 2010) maritime conflict between China and Japan in which Japanese officials claimed that China held up rare earth shipments to Japan (denied by Chinese officials) has heightened the urgency among many buyers to seek diversity in its sources of rare earth materials.

Rare Earth Legislation in the 111th Congress

Congress is concerned about the potential problems with access to rare earth raw materials and downstream products used in many national security applications and clean energy technologies. The Senate Energy and Natural Resources Committee Subcommittee on Energy held a hearing to address the role of strategic minerals and rare earth elements in renewable energy technologies on September 30, 2010. DOE Assistant Secretary for Policy and International Affairs, David Sandalow, discussed Department of Energy efforts to develop a strategic plan for rare earth elements. Legislative proposals have been introduced in the 111th Congress to address these issues.

H.R. 6160, Rare Earths and Critical Materials Revitalization Act of 2010

House bill **H.R. 6160** was introduced on September 22, 2010, and reported (H.Rept. 111-644) by the House Committee on Science and Technology on September 28, 2010. The House approved the bill on September 29, 2010, by a vote of 325-98.

The bill would establish an R&D program within the DOE to assure long-term supply of rare earth materials. The R&D program would, among other things, seek to identify and test potential substitutes, improve extraction, processing, recovery, and recycling technology of rare earth materials. The Secretary of Energy would establish an R&D Information Center and collaborate with members of the European Commission to coordinate activities of mutual interest. The proposed R&D Information Center would be a repository for scientific and technical data, assist scientists and engineers in using the Center, provide advice to the Secretary on the R&D program and promote information sharing among the interested parties. The Secretary of Energy would present a plan to Congress that describes R&D activities and their anticipated contribution to providing rare earth materials to the U.S. economy, explain the requirements of the DOE loan guarantee program and the status of the programs receiving loan guarantee support. After four years the program would be assessed by the National Academy of Sciences (NAS). The program would be authorized for $70 million over a five-year period (2011-2015).

H.R. 6160 would establish a Rare Earth Materials Loan Guarantee program for commercial application of new and improved technologies for the separation, and recovery of rare earths, the preparation of rare earths (i.e., oxides, metal, and alloys) and the application of rare earths in the production of magnets, batteries, optical systems, and electronics, among other things. The Secretary would be required to cooperate with the private sector to assure complete rare earth materials production capacity five years after H.R. 6160 would be enacted into law. The authority to enter into loan guarantees would expire in 2018. The bill would also amend sections 3, 4, and 5 of the National Materials and Minerals Policy Research and Development Act of 1980.

H.R. 4866, the Rare Earths Supply-Chain Technology and Resources Transformation Act of 2010

The House bill, **H.R. 4866** (Coffman) was introduced on March 29, 2010, has 16 co-sponsors, and was referred to the House Committee on Ways and Means, Subcommittee on Trade. The purpose of the bill is

> To reestablish a competitive domestic rare earths minerals production industry; a domestic rare earth processing, refining, purification, and metals production industry; a domestic rare earth metals alloying industry; and a domestic rare earth based magnet production industry and supply chain in the United States.

The bill would establish executive agents (at the Assistant Secretary level) from the Departments of Commerce, Defense, Energy, Interior, and State to serve on an interagency working group. The U.S. Trade Representative (USTR) and White House Office of Science and Technology Policy would also appoint representatives to the working group. The secretaries and the representatives appointed from above agencies would assess rare earth supply chain to determine which REEs are critical to national and economic security. Based on a critical designation, rare earth elements would be stockpiled in the national stockpile administered by the Defense Logistics Agency as part of the National Defense Stockpile. The DLA would, if necessary, make a commitment to purchase rare earth raw materials to process and refine raw materials including purchases made from China if necessary. Stockpiling would be terminated when working group agencies determine REEs are no longer critical to U.S. national security or economic well-being.

The USTR would review international trade practices of REE producers to examine possible market manipulation. Action before the WTO would be possible. Loan guarantees would be provided for supply chain development and the House bill would require a report that would describe "mechanisms" for obtaining loan guarantees for new supply chain development in the U.S. The Department of Defense (DOD) would issue guidance for obtaining loans for new defense supply chains and the Department of Energy (DOE) would issue guidance for development of a domestic supply chain for civilian and commercial purposes.

There is a Sense of Congress statement that would use the Defense Production Act (DPA) of 1950 to develop domestic rare earth supply chain and to provide workforce development and training to reestablish the United States as the preeminent center for rare earths production. R&D funding would also be authorized and if no projects are taking place under DPA, the Congress would want to be notified.

S. 3521, Rare Earths Supply Technology and Resources Transformation Act of 2010

A Senate proposal, **S. 3521**(Murkowski), similar to H.R. 4866, would expedite permitting to increase domestic exploration and development of REEs. A Rare Earth Policy Task Force would be established and would be composed of the Secretaries of Interior, Energy, Defense, Commerce, State, and Agriculture. A representative from OMB, CEQ, and others as deemed necessary by the Secretary of the Interior would also serve on the task force. The Rare Earth Policy Task Force would monitor the acceleration of REE projects, review policies that discourage REE development, and report to Congress on results annually. The Secretaries of Interior, Energy, and others would assess supply chain vulnerability and determine elements "critical" to clean energy technologies. They would prepare a stockpile report to determine if rare earth materials critical for clean energy technology, national and economic security should be stockpiled and determine if legal authorities exist to procure stockpile.

If a stockpile were established for clean energy, national defense, and economic security of the United States, it would contain rare earth oxides, and other storable forms of rare earths and alloys. DOE would issue guidance on obtaining loan guarantees to support a domestic supply chain for clean energy and defense technologies. The DPA would be used to reestablish a domestic supply chain of REEs. A sense of the Congress provision would fund workforce development and training and R&D.

H.R. 5136, the Fiscal Year 2011 National Defense Authorization Act

The House-passed bill (H.R. 5136) would require the Secretary of Defense to assess the rare earth material supply chain to determine if any of the materials were strategic or critical to national security. If the material is determined to be strategic, the Secretary would be required to develop a plan to ensure long-term availability by December 31, 2015. The Secretary shall submit a report to Congress on the assessment and the plan not later than 180 days after enactment of this legislation.

Also, based on congressional findings, among other things, there is an urgent need to eliminate U.S. vulnerability related to the supply of neodymium iron boron magnets and to restore the domestic capacity to manufacture sintered neodymium iron boron magnets used in defense applications. Within 90 days of enactment of this bill the Secretary of Defense would be required to submit a plan, to the appropriate congressional committees, to establish a domestic source of sintered neodymium iron boron magnets used in defense applications.

P.L. 111-84, the Fiscal Year 2010 National Defense Authorization Act

In the proposed House and Senate (H.R. 2647/S. 1390) versions of the defense authorization bill for 2010, Representative Mike Coffman and Senator Evan Bayh introduced legislation to direct the Comptroller General to determine the extent to which specific military weapons systems are currently dependent upon rare-earth materials and the degree to which the United States is dependent upon sources that could be interrupted or disrupted. The measure also directed DOD to describe the risks (both current and projected) involved in the United States' dependence on foreign sources of these materials, and any steps DOD has taken or plans to take to address any potential risks to national security.[30] The measure was passed in the Fiscal Year 2010 National Defense Authorization Act.[31]

Possible Policy Options

Authorize and Appropriate Funding for a USGS Assessment

In addition to the two current legislative proposals, Congress could authorize and appropriate funding for a USGS comprehensive assessment to identify economically exploitable REE deposits (as a main product or co-product), and where REE could be exploited as a byproduct. Authorizations and additional appropriations could be supported for basic science research on substitutes and efforts at secondary recovery for REEs. Additionally, R&D may be necessary on how to proceed in the exploitation of high-thorium monazite deposits where REE could be produced as a byproduct.

Support and Encourage Greater Exploration for REE

Supporting/encouraging greater exploration for REE efforts in the United States, Australia, Africa, and Canada could be part of a broad international strategy. There are only a few companies in the world that can provide the exploration and development skills and technology for REE development. These few companies are located primarily in Canada, Australia, China, South Africa, and the United States, and may form joint ventures or other types of alliances for R&D, and for exploration and development of REE deposits worldwide, including those in the United States. Whether there should be restrictions on these efforts in the United States is a question that Congress may ultimately choose to address.

Challenge China on Its Export Policy

Challenging China on its export restrictions through the WTO would involve filing a dispute based on WTO rules that generally prohibit members from imposing restrictions (i.e., quotas) or other restraints (e.g., minimum prices or licensing) on exports. In June 2009, the United States filed a dispute over raw material exports from China, which included: bauxite, coke, fluorspar, magnesium, manganese, silicon carbide, silicon metal, yellow phosphorus and zinc.[32] Some REE analysts assert that China sets export restrictions to meet growing Chinese demand for raw materials and to force the manufacturing of end-use products in China.[33]

Establish a Stockpile

Establishing a government-run economic stockpile and/or private-sector stockpiles that would contain supplies of specific REE broadly needed for "green initiatives" and defense applications is a policy advocated by some in industry and government. This may be a prudent investment. Generally, stockpiles and stockpile releases could have an impact on prices and supply but would also ensure supplies of REE materials (oxides, metals, etc.) during times of normal supply bottlenecks. An economic stockpile could be costly and risky, as prices and technology may change the composition of REEs that are needed in the economy.

According to USGS,[34] DOD along with USGS is examining which of the REEs might be necessary in the National Defense Stockpile (NDS). In the recent past, NDS materials were stored for wartime use based on a three-year war scenario. Some of the rare earth elements contained in the National Defense Stockpile were sold off by 1998. However, rare earth elements were never classified as strategic minerals.[35] DOD had stockpiled some yttrium but has since sold it off, and none of the REEs have been classified as strategic materials. A critical question for stockpile development would be: What materials along the supply chain should be stockpiled? For example, should the stockpile contain rare earth oxides or alloyed magnets which contain the REEs, or some combination of products?

The National Research Council (NRC) has produced an in-depth report on minerals critical to the U.S. economy and offers its analysis as described here: "... most critical minerals are both essential in use (difficult to substitute for) and prone to supply restrictions."[36] While the NRC report is based on several availability criteria used to rank minerals for criticality (geological, technical, environmental and social, political, and economic), REEs were determined to be critical materials assessed at a high supply risk and the possibility of severe impacts if supplies were restricted. Some of the REE applications are viewed as more important than others and some are at greater risk than others, namely the Heavy Rare Earth Elements (HREEs), as substitutes are unavailable or not as effective.[37]

The federal government and private sectors are beginning to address how to secure reliable rare earth materials (raw materials through metals and alloys) from China and non-Chinese sources in the short term, and how to rebuild the U.S. supply chain for the long term.

End Notes

[1] U. S. mineral policies provide a framework for the development of domestic metal mineral resources and for securing supplies from foreign sources. Specifically, the Mining and Minerals Policy Act of 1970 (30 U.S.C. §21a) declared that it is in the national interest of the United States to foster the development of the domestic mining industry "... including the use of recycling and scrap." The National Materials and Minerals Policy, Research and Development Act of 1980 (30 U.S.C. 1601), among other things, declares that it is the continuing policy of the United States to promote an adequate and stable supply of materials necessary to maintain national security, economic well-being and industrial production, with appropriate attention to a long-term balance between resource production, energy use, a healthy environment, natural resources conservation, and social needs.

[2] U.S. Department of the Interior (DOI), Geological Survey (USGS), *Minerals Yearbook, Volume 1*, 2007, Rare Earths (Advance Release).

[3] DOI/USGS, *Rare Earth Elements—Critical Resources for High Technology*, Fact Sheet 087-02.

[4] Rare Earth Minerals: The Indispensable Resource for Clean Energy Technologies, presented at Minerals for a Green Society Conference, Washington, DC, by Mark A. Smith, CEO, Molycorp Minerals, February 4, 2010.

[5] DOI/USGS Minerals Yearbook, Volume 1, 2007.

[6] This section was prepared by Valerie Grasso, CRS Foreign Affairs, Defense, and Trade Division.

[7] The following information was accessed on March 10, 2010, from the Defense National Stockpile Center website, at https://www.dnsc.dla.mil/inside.asp., and GlobalSecurity.org, at http://www.globalsecurity.org/military dnsc.htm. "The National Stockpile operates under authority of the Strategic and Critical Materials Stockpiling Act (50 U.S.C. 98-h-2(a)). This act provides that strategic and critical materials are stockpiled in the interest of national defense to preclude a dangerous and costly dependence upon foreign sources of supply in times of national emergency. The Defense National Stockpile Center administers the storage, management, and disposal of the Nation's inventory of strategic and critical materials essential to the military and industrial requirements of the United States in times of national emergency. The Congress of the United States has

authorized the Defense National Stockpile Center to sell commodities that are excess to Department of Defense needs. Since 1993, DNSC sales have totaled approximately $6.6 billion."

[8] For further discussion on the Berry Amendment and the Specialty Metals provision, see CRS Report RL3 1236, *The Berry Amendment: Requiring Defense Procurement to Come from Domestic Sources*, and CRS Report RL33751, *The Specialty Metal Provision and the Berry Amendment: Issues for Congress*, both by Valerie Bailey Grasso.

[9] For further discussion on precision-guided munitions, see CRS Report RL33539, *Intelligence Issues for Congress*, by Richard A. Best Jr., and CRS Report RL30727, *Airborne Intelligence, Surveillance, and Reconnaissance (ISR): The U2 Aircraft and Global Hawk UAV Programs*, by Richard A. Best Jr. and Christopher Bolkcom.

[10] Haxel, Gordon B, Hendricks, James B., and Oris, Greta J. Rare Earth Elements: Critical Resources for High Technology. *U.S. Geological Survey*, Fact Sheet 087-82, accessed online on March 10, 2010, at http://pubs.usgs.gov/fs/ 2002/fs087-02/.

[11] Green Jeffrey A. Defense, Energy Markets Should Brace for Shortages of Key Materials. *National Defense Industrial Association*, October 2009; U.S. Lacks Data on Supply of Minerals Critical to Economy, National Security; Defense Stockpile is Ineffective. National Academy of Sciences, October 5, 2007, accessed online at http://www8.nationalacademies.org/onpinews/newsitem.aspx?RecordID=10052007.

[12] Chang, Gordon G. Why Does China Want BP? *Forbes.com,* July 14, 2010. http://www.forbes.com/ 2010/07/14/bpchina-oil

[13] Magnet Materials Supply Chain Players Propose Six-Point Plan to Address Impending Rare Earths Crisis, *USMMA,* February 4, 2010, accessed online at http://www.usmagnetmaterials.com/?p=74.

[14] Ibid.

[15] Bastnaesite is mineral with the formula (Ce, La)CO3(F,OH) that may contain other rare earth elements.

[16] Monazite is a mineral with the formula (Ce, La, Nd, Th)PO_4 that may contain other rare earth elements.

[17] DOI/USGS Circular 930 N, *International Strategic Minerals Inventory Summary Report—Rare Earth Oxides,* by Wayne Jackson and Grey Christiansen, 1993.

[18] Dr. Anthony N. Mariano, *The Nature of Economic REE and Y Minerals on a World Level*, presented at the MIT Energy Initiative Workshop, April 29, 2010.

[19] Ibid.

[20] The Jack Lifton Report, *The Rare Earth Crisis—Part I*, by Jack Lifton, October 2009.

[21] *Rare Earth Strategic Supplies More Important Than Price,* Industrial Metals/Minerals Interview of Jack Lifton by The Gold Report, December 14, 2009.

[22] Ibid.

[23] Ibid.

[24] "China Cuts Rare Earth Export Quota," Bloomberg News Report, July 9, 2010.

[25] Jack Lifton, "Is The Rare Earth Supply Crisis Due to Peak Production Capability or Capacity," michaelperelman.worldpress.com, September 6, 2009.

[26] U.S. Government Accountability Office, *Rare Earth Materials in the Defense Supply Chain*, GAO-10-617R, April 1, 2010.

[27] Op. Cit., Lifton Interview by The Gold Report, December 14, 2009.

[28] China's Rare Earth Elements Industry: What Can the West Learn? by Cindy Hurst, Institute for the Analysis of Global Security, March 2010.

[29] Ibid.

[30] McCormack, Richard. China's Complete Control of Global High-Tech Magnet Industry Raises U.S. National Security Alarms. *Manufacturing & Technology News*, Vol. 16, No. 16, September 30, 2009, viewed online at http://www.manufacturingnews.com/news/09/0930/magnets

[31] P.L. 111-84 was signed into law on October 28, 2009. The text, as it appears in the bill, reads as follows.

Sec. 843. Report on Rare Earth Materials in the Defense Supply Chain.

(a) Report Required.—Not later than April 1, 2010, the Comptroller General shall submit to the Committees on Armed Services of the Senate and House of Representatives a report on rare earth materials in the supply chain of the Department of Defense.

(b) Matters Addressed.—The report required by subsection (a) shall address, at a minimum, the following: (1) An analysis of the current and projected domestic and worldwide availability of rare earths for use in defense systems, including an analysis of projected availability of these materials in the export market. (2) An analysis of actions or events outside the control of the Government of the United States that could restrict the access of the Department of Defense to rare earth materials, such as past procurements and attempted procurements of rare earth mines and mineral rights. (3) A determination as to which defense systems are currently dependent on, or projected to become dependent on, rare earth materials, particularly neodymium iron boron magnets, whose supply could be restricted—(A) by actions or events identified pursuant to paragraph (2); or (B) by other actions or events outside the control of the Government of the United States. (4) The risk to national security, if any, of the dependencies (current or projected) identified pursuant to paragraph (3). (5) Any steps that the Department of Defense has taken or is planning to take to address any such risk to national security. (6) Such recommendations for further action to address the matters covered by the report as the Comptroller

General considers appropriate. (c) Definitions.—In this section: (1) The term "rare earth" means the chemical elements, all metals, beginning with lanthanum, atomic number 57, and including all of the natural chemical elements in the periodic table following lanthanum up to and including lutetium, element number 71. The term also includes the elements yttrium and scandium. (2) The term "rare earth material" includes rare earth ores, semi-finished rare earth products, and components containing rare earth materials.

[32] Office of the U.S. Trade Representative, Press Release, "WTO Case Challenging China's Export Restraints on Raw Material Inputs," June 23, 2009.

[33] Irma Venter, "Investors take closer look at rare earth elements as technology, green revolution pick up pace," Mining Weekly Online, http://www.miningweekly.com, September 18, 2009.

[34] Phone interview with Jim Hedrick, Rare Earth Specialist, USGS, October 1, 2009.

[35] For a discussion of the strategic materials for defense uses, see CRS Report RL33751, *The Specialty Metal Provision and the Berry Amendment: Issues for Congress*, by Valerie Bailey Grasso.

[36] National Research Council, *Minerals, Critical Minerals, and the U.S. Economy*, National Academies Press, 2008.

[37] DOI/USGS, Minerals Yearbook, Volume 1, 2007.

CHAPTER SOURCES

The following chapters have been previously published:

Chapter 1 – This is an edited reformatted and augmented version of a United States Department of Energy publication, dated December 2010.

Chapter 2 – This is an edited reformatted and augmented version of a Congressional Research Service publication, report R41347, dated September 30, 2010.

INDEX

D

E

F

G

H

N

O

P

Q

R

S

T

U

V

W

Y

Z